# Clean Disruption of Energy and Transportation

How Silicon Valley Will Make Oil, Nuclear,
Natural Gas, Coal, Electric Utilities
and Conventional Cars Obsolete by 2030

# Clean Disruption of Energy and Transportation

How Silicon Valley Will Make Oil, Nuclear, Natural Gas, Coal, Electric Utilities and Conventional Cars Obsolete by 2030

## TONY SEBA

Clean Planet Ventures
Silicon Valley, California, USA

First Beta Edition v.0.000.04.28.14

June 15, 2014

Printed in the United States of America.
First Beta Edition

ISBN-13 978-0-692-21053-6
For information about quantity discounts or about permission to
reproduce selections from this book, Email: info@tonyseba.com
www.tonyseba.com

Cover Art ©2014 by Tony Seba.
Many of the product names referred to herein are trademarks or
registered trademarks of their respective owners.

Library of Congress Cataloging-in-Publication Data

Seba, Tony
Clean Disruption of Energy and Transportation: How Silicon Valley
Make Oil, Nuclear, Natural Gas, and Coal Obsolete by 2030
– 1st ed. p. cm.

Includes bibliographical references and index.
ISBN-13 978-0-692-21053-6

# To Maylén Rafuls

This book would not have been possible without you.
Thanks for your support.

# Table Of Contents

# Acknowledgements

**"Victory has a hundred fathers and defeat is an orphan."**- President John F. Kennedy

I want to thank all who agreed to be guest speakers in my "Clean Energy and Transportation—Market and Investment Opportunities" and "Understanding and Leading Market Disruption" courses at Stanford University, as well as those who agreed to be interviewed for this book: Masato Inoue (Nissan), Takeshi Mitamura and Kimihiko Iwamura (Nissan Research Center Silicon Valley), Danny Kennedy (Sungevity), David Arfin, (SolarCity), Kevin Smith (SolarReserve), Jose Martin (Sener USA), Craig Horne (Enervault), Peter LeLievre (Chromasun), Raj Atluru (DFJ Ventures), Manny Hernandez (SunPower), G.G.Pique (Energy Recovery), Peter Childers (Utility Scale Solar), Steve Nasiri (Invense), Abe Reichental (3D Systems), Rich Mahoney (SRI International), Andreas Raptopoulos (Matternet), Dan Rosen (Mosaic), and Emily Kirsch (SFUNCube).

I want to thank my students in my Stanford courses "Clean Energy and Transportation—Market and Investment Opportunities" and "Understanding and Leading Market Disruption." The intersection of my clean energy class and my disruption class provided the core ideas for this book. I am blessed to be able to teach the innovators and entrepreneurs who are changing the world in a positive direction. Many of them have gone on to start or join Silicon Valley disruptive companies in clean energy, clean transportation, and information technology. They have pushed the boundaries of technology, business models, and product innovation. Some have gone on to develop hundreds of megawatts of solar and wind power. Others have helped make a positive impact at think tanks, non-governmental organizations, and public policy institutions.

I do my best to educate, inspire, and push their thinking beyond previous perceptions and boundaries. They, in turn, challenge me and inspire me to deliver an exceptional experience that lasts a lifetime. I thank them for that.

I want to thank Hal Louchheim for giving me the opportunity to teach at Stanford twelve years ago. I have created and taught five different courses at Stanford, and Hal has always been trusting and supportive of my efforts. I also want to thank Dan Colman, head of Stanford's Continuing Studies, and his wonderful staff.

I stand on the shoulders of countless scientists, engineers, and entrepreneurs who have built solar and wind power technologies, electric vehicles, autonomous cars, electric energy storage devices, robotics, intelligent devices, sensors, artificial intelligence, and myriad other technologies to enable the products, services, and business models that can bring about a clean energy and clean transportation world. They have paved the way for the rest of us to enjoy not just a clean future, but a more democratic world.

I want to thank Elena Castanon and her wonderful team at Wikreate. They were the creative designers behind this book and my new website (tonyseba. com) and elevated both to a higher level. Thanks to Peter Weverka, who edited my manuscript and clearly turned it into a more readable book. Thanks to Joe Deely for his feedback.

I want to thank the staff at Village Market Coffee in San Francisco. A friend of mine said to write a book "all you need is love and the Internet." I'd add coffee to that mix.

I want to thank Bhavesh Singh (Aegis Capital Partners) for your friendship and for being a great business partner.

Finally, I want to thank Maylén Rafuls for all your support. This book would not have happened without you. You are the best.

To all of you, thanks for who you are and what you do.

# Introduction:
# Energy and
# the Stone Age

*"An age is called Dark not because the light fails to shine,*

*but because people refuse to see it."*

*- James Michener.*

*"If the rate of change on the outside is greater than*

*the rate of change on the inside, the end is near."*

*- Jack Welch, former CEO, General Electric*

*"It always seems impossible until it's done."*

*- Nelson Mandela*

The Stone Age did not end because humankind ran out of stones. It ended because rocks were disrupted by a superior technology: bronze. Stones didn't just disappear. They just became obsolete for tool-making purposes in the Bronze Age.

The horse and carriage era did not end because we ran out of horses. It ended because horse transportation was disrupted by a superior technology, the internal combustion engine, and a new, disruptive 20th century business model. Horses didn't just disappear. They became obsolete for the purposes of mass transportation.

The age of centralized, command-and-control, extraction-resource-based energy sources (oil, gas, coal and nuclear) will not end because we run out of petroleum, natural gas, coal, or uranium. It will end because these energy sources, the business models they employ, and the products that sustain them will be disrupted by superior technologies, product architectures, and business models. Compelling new technologies such as solar, wind, electric vehicles, and autonomous (self-driving) cars will disrupt and sweep away the energy industry as we know it.

The same Silicon Valley ecosystem that created bit-based technologies that have disrupted atom-based industries is now creating bit- and electron-based technologies that will disrupt atom-based energy industries.

## Clean Disruption of Energy and Transportation.

The industrial era of energy and transportation is giving way to an information technology and knowledge-based energy and transportation era. The combination of bit-based and electron-based technologies will put an end to conventional atom-based energy and transportation industries. The disruption will be a clean one and have the following characteristics:

### 1. Technology-based disruption.
The clean disruption is about digital (bit) and clean energy (electron) technologies disrupting resource-based (atom-based) industries. Clean energy (solar and wind) is free. Clean transportation is electric and uses clean energy derived from the sun and wind. The key to the disruption of energy lies in the exponential cost and performance improvement of technologies that convert, manage, store, and share clean energy. The clean disruption is also about software and business model innovation.

### 2. Flipping the architecture of energy.
Just as the Internet and the cell phone turned the architecture of information upside-down, the clean disruption will create an energy archi-

tecture that is different from the one we know today. The new energy architecture will be distributed, mobile, intelligent, and participatory. It will overturn the existing energy architecture, which is centralized, command-and-control oriented, secretive, and extractive. The conventional energy model is about Big Banks financing Big Energy to build Big Power Plants or refineries in a few selected places. The new architecture is about everyone financing everyone to build smaller, distributed power plants everywhere.

### 3. Abundant, cheap, and participatory energy.
The clean disruption will be about abundant, cheap, and participatory energy. The existing energy business model is based on scarcity, depletion, and command-and-control monopolies. The clean disruption is similar to the information technology revolution that overturned the old publishing and information model and made information abundant, participatory, and essentially free.

### 4. Clean disruption is inevitable.
The clean disruption of energy and transportation is inevitable when you consider the exponential cost improvement of disrupting technologies; the creation of new business models; the democratization of generation, finance, and access; and the exponential market growth.

### 5. Clean disruption will be swift.
It will be over by 2030. Maybe before.

Oil, natural gas (methane), coal, and uranium will simply become obsolete for the purposes of generating significant amounts of electricity and powering the automobile. These energy sources will still have uses. For example, uranium will be used to make nuclear weapons and natural gas will be used for cooking and producing fertilizer. Obsolescence and clean disruption will not put an end to incumbent industries. We still have vinyl records, sailboats and jukeboxes. These niche market products will survive, but energy and transportation will not be the multi-trillion dollar energy heavyweights that they are today.

In twenty years we'll wonder how we put up with the horrendous consequences of the incumbent, conventional, $8 trillion-a-year energy industry. If Nikola Tesla and Thomas Alva Edison rose from the dead, they would recognize the industry that they helped build a century ago — and they would be disappointed at how little it has changed. Today's versions of Tesla and Edison are creating technologies, products, and business models that will dismantle the extractive, centralized, dirty-energy age in which we live.

The first wave of energy disruption has already begun with distributed solar and wind generation. It won't be long before the next wave crashes over the remains of the first one.

Transportation is a $4 trillion industry globally. The transportation industry is inextricably linked with energy. As this book explains, the internal combustion engine automobile will soon be disrupted, an event which will, in turn, send disruptive shockwaves through the oil industry.

The first wave of disruption of the century-old automotive industry is well underway with electric vehicles. The second disruptive wave, the self-driving car, will hit before the first wave is finished crashing. Transportation will never be the same again.

This book is about how a new technology-based infrastructure and a set of products and services governed by the economics that have made Silicon Valley a source of market disruption over the last generation will disrupt energy industries that have barely evolved over the past hundred years.

## A Classic Silicon Valley Technology Disruption

Companies such as Apple, Google, Intel, Cisco, Facebook, Twitter, and eBay are governed by information economics. These technology companies grew fast and strong because of the economics of increasing returns.

Resource-based energy companies are based on the economics of decreasing returns. Silicon Valley is about abundance, business model innovation, participatory culture, and democratizing power. Resource energy is about scarcity, extractive thinking, hierarchical culture and centralized power.

To explain the power of clean disruption, it helps to look at a recent industry that was disrupted by Silicon Valley — film photography.

## Zero Marginal Cost and Waves of Disruption

The age of film photography did not end because we ran out of film. We did not run out of any of the components needed to make film or film cameras. Film photography was destroyed by rapid improvements in digital imaging and information technologies, disruptive business models, and a participatory culture with which industry leaders Kodak and Fujifilm simply could not compete.

Twentieth-century photography leader Kodak's business model was to make money every single time anyone, anywhere clicked his or her camera.

Every click was a cash transaction for Kodak. Every click involved burning film (cash for Kodak). The film had to be processed with special paper (cash for Kodak.)

The paper needed a printer developed especially for after-market photo stores (cash for Kodak). Want to double size it and get two copies of each photo? Cash for Kodak.

Digital cameras changed the equation. Once a photographer had a digital camera the marginal cost of taking additional pictures dropped essentially to zero. The photographer did not have to pay for film, film processing, or printing photographs. Just load the files on your computer and enjoy. Erase your camera USB drive, take as many pictures as you like, load the files on your computer and enjoy. Repeat forever.

The energy and transportation industries have a business model similar to Kodak's. Every time you flip a switch to turn on a light, more cash is paid to the utility. Every flip of the switch involves burning coal, oil, gas, or uranium and, again more cash for resource-based energy suppliers. Every time you press the gas pedal in your car, you give cash to the oil industry. Substituting natural gas or ethanol for gasoline doesn't change the business model. Every time you press the gas pedal you still burn fuel and give cash to the energy industry.

Solar and wind power change the energy equation in the same way that digital cameras changed the film camera equation. After you build a solar rooftop installation, the marginal cost of each additional unit of energy drops essentially to zero because the sun and the wind are free. Flipping a light switch burns nothing and means zero cash for the utility.

This applies not just to rooftop solar. Utility scale solar and wind also change the equation in competitive wholesale electricity markets. Solar and wind have a marginal cost of zero. Chapter 3 explains how zero marginal costs are already disrupting utilities that rely on coal, nuclear, gas, and oil for energy generation.

Kodak and its film photography supply chain did try to compete with digital photography (see Figure I-1). For instance, Kodak developed technologies to speed up development time from days to hours. But what was disruptive about digital photography was not just the technology itself, but the business model innovations that came with it. Under these models, the marginal cost of a new photo dropped to zero. That was something Kodak could not compete with.

*Figure I.1 –Obsolete one-hour film developing store. (Photo: Tony Seba)*

The story of how digital photography disrupted traditional photography didn't end with Kodak. The next disruption wave came in the form of San Francisco-based Flickr, which made it easy to publish and share photos online. Again the cost of uploading and storing each picture dropped to zero. Companies like Picasa made it easy to store photos on your computer and online. Again, the cost of each additional picture was nil.

Next came the social media disruption wave. Facebook became the largest photo publisher in the world.

Not much later, the smartphone disruption wave swooped in. Smartphone cameras were just as good as standalone cameras or at least good enough for everyday photo taking. You could take photos, process them, and instantly publish them online without leaving your smartphone. Instagram, a startup company in San Francisco with a dozen employees, simplified this process and within a few months became the fastest growing photo publisher in the world. Facebook acquired Instagram for $1 billion before Instagram could become an existential threat to the social network.

What happened in photography and what is happening in many other industries is what I call "waves of disruption" or "disruptive waves." These waves used to happen every century or maybe every generation. The computer industry sped up disruption waves so they occur every decade or so (see Figure I.2).

Now we live in an era of permanent disruption. Just as soon as the disrupting companies start celebrating their triumph over the former incumbents, they become the targets of the next wave of disruptors.

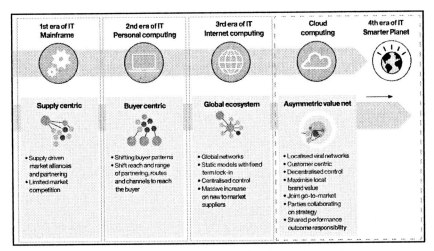

*Figure I.2—Disruptive waves in information technology. (Source: IBM)* [1]

As soon as Sony started celebrating the triumph of its digital cameras over Kodak and Fuji, it was commoditized by web photo companies like Flickr. Flickr, meanwhile, was acquired by Yahoo!, and while the bottles of bubbly were being imbibed by Flickr employees, the Flickr photo site was disrupted by social media hubs like Facebook. Facebook, in turn, was threatened with disruption by Instagram. Now Instagram and Facebook are being threatened by another fast-growing wave led by SnapChat.

The disruption of energy and transportation as we know it today is being led by three main sets of technology-based products:
  1. Solar
  2. Electric vehicles
  3. Autonomous (self-driving) cars

Solar is on its way to disrupting all forms of conventional energy. Solar is already cheaper than nuclear. It's already cheaper than retail electricity in hundreds of markets around the world, from Berlin to Seville to Palo Alto. In some markets solar has already pushed wholesale electricity prices down by as much as 40 percent.

Solar photovoltaic (PV) companies have decreased their costs by a factor of 154, a classic technology cost curve. Technology companies have an unparalleled record of lowering costs exponentially while increasing quality exponentially. The same economics that governed digital cameras, disk drives, microprocessors, routers, and mobile phones now govern solar PV technology development.

The electric vehicle is already better, faster, and safer than the internal combustion engine (gasoline) vehicle. Electric vehicles (EVs) are also cheaper to operate and maintain. An electric vehicle is still more expensive to purchase upfront, mainly due to battery costs. However, like other technology products, the technology cost curve of EVs points to a disruption soon; innovative business models will only accelerate the transition from gasoline vehicles to electric vehicles.

Internal combustion engine car companies will have their Kodak moment sooner than they think. By 2025, gasoline engine cars will be unable to compete with electric vehicles.

The autonomous (self-driving) vehicle will soon be better, faster, cheaper, and safer than vehicles driven by human drivers. The disruptive wave brought about by self-driving cars will wipe the last vestiges of the gasoline car and oil industries.

## Technology Convergence and the Clean Disruption

In the clean energy field, the disruptors (solar, electric vehicles, and autonomous cars) complement and accelerate one another's adoption. For this reason, the disruption of energy and transportation as we know them today will be a dynamic one.

Think of the cell phone, the personal computer, and the Internet. They started out as different sets of products serving different markets, but their symbiosis complemented and accelerated one another's adoption in the marketplace. Cell phone, computer, and Internet router providers all benefited from the increased investment and R&D in smaller, more powerful, modular energy-efficient microprocessors, graphics processors, data storage, and connectivity. In the end these formerly disparate industries converged. Together they formed a massive mobile computing infrastructure. This infrastructure encompasses everything from cell phones, smart phones, tablet computers and personal computers to data centers that host cloud-computing services.

These technologies have disrupted century-old industries while improving the lives of billions of people around the world.

Disruption comes in waves and we're still seeing disruption within the different classes of computers. Sales of personal computers are down and sales of mobile Internet platforms like the smartphone and tablets are up. Transitions in technology markets can be swift. It took twelve years to reach 50 million laptops, seven years to reach 50 million smartphones, but only two years to reach 50 million tablets.[2]

Microsoft's Windows PC is out while Apple iOS iPhones and Google Android smartphones are in. Apple's iPad tablet is in and everyone else is still trying to catch up with Apple. The percentage of YouTube traffic that comes from mobile Internet went from 6 percent in 2011 to 25 percent in 2012 to 40 percent in 2013.[3]

Similarly solar, the electric vehicle, and the autonomous vehicle started out as different sets of products and markets, but their symbiosis will complement and accelerate one another's technological development and adoption in the marketplace.

Increasing investments in electricity storage technologies in the automotive industry have led to more innovation and a subsequent drop in the cost of batteries like Lithium-Ion. As Li-on batteries become cheaper, they can increasingly be used — and economically be used — for solar and wind energy storage. The increased demand from solar and wind increases the scale of existing Li-on providers, which in turn pushes down the cost of EVs, solar, and wind.

The increasing demand for electric vehicles and solar will attract even more investment in these technologies. Innovative companies that can invent new ways to push costs down and push quality up will thrive. This virtuous cycle of increasing demand, increasing investment, and increasing innovation will dramatically lower costs; it will exponentially improve the quality benefits to both the clean energy and clean transportation industries; it will also lead to a convergence in which batteries can be used for transportation and for grid storage. Electric vehicles can be charged at work and become a source as well as a user of energy for the home. The result will be a swift transition from liquid-energy transportation to electric transportation.

The self-driving car will benefit from improvements in technologies such as artificial intelligence, sensors, graphics processing, robotics, broadband wireless communications, advanced materials, 3D visualization, Lidar, and 3D printing. These technologies will also benefit solar, wind, and electric vehicles.

Today, the self-driving Google car uses advances in Lidar 3D visualization technology (see Chapter 5). Lidar can also be used to make high-resolution maps for use in forestry, archeology, seismology, and other fields. For instance, the National Oceanic and Atmospheric Administration (NOAA) uses Lidar to collect data and develop 3D shoreline mapping tools. These tools will accurately map and project flooding and storm surges on the coasts of the United States.[4]

Cities everywhere, from Cambridge, Massachusetts to San Diego, California use Lidar to develop "birds-eye view" 3D maps for use in urban planning, architecture, and design. Imagine a 3D Google-Earth-like SimCity map with which you can zoom in to any building, look at buildings from different angles, and virtually design different versions of a new house, clinic, or park.

As it happens, Lidar data developed to protect coastal populations, track earthquake fault-lines, and help city planners with urban design can also be used to develop more accurate assessments of a region's solar potential; Lidar data can even be used to design solar installations on the rooftops of buildings. A recent MIT study concluded that designing a solar installation with Lidar data maps resulted in a "higher prediction of solar PV yield and a 10.8-percent reduction in costs." [5] Lidar can also be used to measure the speed, angle, and intensity of the wind, data that managers can use to improve the planning and operation of wind power plants.

Lidar is an example of an exponentially improving technology that can be used in autonomous and electric vehicles, solar and wind. As the market for autonomous vehicles grows, demand for Lidar will increase, which will attract more research and development investment in Lidar. The combination of these factors will lead to lower costs for Lidar products, which will benefit not just the autonomous vehicle industry but also solar and wind.

Because the self-driving car is basically a mobile computer, it will also benefit from improvements in existing Silicon Valley computing and communications technologies: data storage, computers, operating systems and applications software, communications, and graphics accelerators.

"Electric vehicles are the natural platform for autonomous cars," said Takeshi Mitamura, director of the Nissan Research Center – Silicon Valley, speaking from his office in Sunnyvale. Nissan has announced the launch of an autonomous car by 2020.[6]

**Participatory Energy, Business Model Innovation and Disruption**

The clean disruption of energy and transportation is also about innovative business models, designing new products and services, dominating small markets, and growing exponentially until the incumbents become collateral damage.

This disruption is also about a whole new architecture of energy. The Internet disrupted information, communications, and computing in large part due to its distributed architecture. Information technology shifted from a centralized, supplier-centric, hierarchical model to a distributed, customer-centric, participatory model.

The way we produced, stored, distributed, and consumed information changed radically.

Distributed technologies, which were enabled by innovative, disruptive business models, in turn allowed new technologies to flourish, and these technologies in their turn caused a change in the culture of information. Information changed from a centralized to a participatory model. In the end, the shift from centralized to distributed information technologies changed everything about the industry; it even changed society at large. People don't want to just consume content; they want to create and share it. Companies that enable people to participate in the generation and dissemination of content have been amply rewarded. Witness the rise of Facebook, Twitter, and LinkedIn.

Energy will be no different.

Millions of routers were needed to build and underpin the Internet infrastructure. Today, solar panels and electric vehicles are needed to underpin the new energy infrastructure and transportation infrastructure. But make no mistake, solar and electric vehicles are also about a shift in the architecture of energy. This new energy architecture will change the way energy is produced, stored, distributed, and consumed. It will bring about new disruptive technologies and business models, and even a cultural shift in the way we think about energy.

## The Participatory Energy Model

The information technology revolution pushed processing power and intelligence from the center to the edges. We went from the mainframe, to the minicomputer, to the personal computer, to the cell phone and tablet in less than three decades. The nodes are getting smaller, more connected and more intelligent. We're far from done with this transition. The trillion-sensor world is right around the corner.[7]

The information technology revolution was not brought about only by the miniaturization of technologies. This was a transition from a supplier-centric, centralized information model to a user-centric, participatory information model.

Twenty-first century digital consumers have grown to feel empowered by distributed technologies built on the Internet and the smart mobile phone. Consumers who previously had access to one or two local newspapers now can get information from anywhere in the world. The local paper hasn't died but it's wounded and weak.

Following in the footsteps of information technology, the energy and transportation disruption is quickly moving towards a participatory energy model. We're headed toward a distributed architecture of energy production and usage made possible by software, sensors, artificial intelligence, robotics, smartphones, mobile Internet, big data, analytics, satellites, nanotechnology, electricity storage, materials science, and other exponentially improving technologies.

Solar is causing energy production to be pushed to the edges (customer site) from the center (large, centralized, hub-and-spoke power plants). The nodes are getting smaller, more modular, more connected, and more intelligent.

Welcome to the age of participatory energy, where every end user will be able to contribute to the financing, generation, storage, management, and trading of energy.

Thanks to the distributed nature of solar energy production and the open accessibility of information about energy, consumers can choose where they get their energy. The mobility and connectivity of electric vehicles will turn these vehicles into intelligent energy generation, storage, and management devices. Soon individuals will help decide which vendors provide energy and who will manage its efficient usage.

### The Economics of Silicon Valley Technology: Increasing Returns

Distributed solar generation, the electric vehicle and the autonomous vehicle are information products. As such, they are governed by information economics and increasing returns. They are subject to Moore's Law as much as personal computers and tablets.

## Increasing and Decreasing Returns: Technology vs. Extraction

Conventional energy resource economics is about decreasing returns. For this reason, conventional energy can't compete with technology industries based on increasing returns.

Take the new darling of conventional energy: hydraulic fracturing, also known as "fracking." To "frack" a single oil or gas well requires hundreds of trucks, millions of gallons of water, and tons of sand with hundreds of chemicals blasted through the ground. You also need thousands of miles of pipelines, massive factories to liquefy or compress the gas before it can be shipped or stored, and massive ports with massive plants to decompress the gas and pipe it again to the power plant. Power generation can start only after all this Rube-Goldberg-device-like process is complete.

The returns on these wells start decreasing as soon as you start pumping the oil or gas. Despite all the talk of abundance and a "golden age of energy," fracked wells may deplete by 60 to 70 percent the first year alone.[8]

The industry has started calling this depletion phenomenon the "Red Queen Syndrome" (after the Red Queen in Through The Looking Glass, who tells Alice "it takes all the running you can do just to stay in place"). Because of Red Queen Syndrome, you need to frack millions of new wells just to keep up with existing production. This is not just a "fracking" phenomenon. Production from traditional wells declines by half in about two years, after which the wells drip on for a few more years.

Extraction economics is about decreasing returns:
- The more you pump, the less each well produces.
- The more you pump, the less the neighboring well gets.
- The more you pump, the more each unit of energy will cost in the future.

Solar, electric vehicles and the clean disruption are about increasing returns.

Solar photovoltaic (PV) panels have a learning curve of 22 percent. PV production costs have dropped by 22 percent with every doubling of the infrastructure. The more demand there is in the market, the less your neighbor pays for her panels, and the more your neighbor benefits. Every time a solar power plant is built in Germany, Californians benefit from lower costs when the next solar power plant is built. Every solar panel sold in Australia cuts the cost of the next solar panel in South Africa. Lower costs benefit all new solar customers.

Every large solar power plant in the desert benefits not only the people who buy its power, but everyone who buys solar power in the future.

The higher the demand for solar PV, the lower the cost of solar for everyone, everywhere. Your neighbor benefits, the warehouse owner in Australia benefits, and future buyers of solar benefit from lower costs. All this enables more growth in the solar marketplace, which, because of the solar learning curve, further pushes down costs.

This mutually beneficial arrangement is the opposite of extraction industries like oil and gas. When China's demand for oil surged in the last decade, world prices for oil went up by a factor of ten. The higher the demand for oil in Beijing, the higher gasoline prices are in Palo Alto and Sydney.

This is not just a theoretical framework. Solar PV has improved its cost basis by more than five thousand times relative to oil since 1970 (see Chapter 7). By 2020, as the market for solar expands, solar will improve its cost basis relative to oil by twelve thousand times (see Chapter 7).

The economics of energy resource extraction, based on decreasing returns, just cannot compete with the economics of technology industries and its increasing returns.

The Red Queen Syndrome pushes the fossil fuel industry not just to extract more wells but to dig deeper, use harsher chemicals, and create more waste-lands. The fossil fuel industry has to do this just to stay in place. The BP Gulf Oil disaster and the monstrosity of the Alberta Oil Sands are not exceptions; they are the inevitable result of The Red Queen having to "run harder just to stay in place."

### Network Effects and the Clean Disruption of Energy and Transportation

Network effects explain why the value of a network increases exponentially even when adoption increases linearly. Network effects are the reason AT&T so thoroughly dominated telephony in the U.S. for a century; they explain why Microsoft Windows has generated so much cash for three decades and why Apple's iOS and Google's Android platforms have become so valuable. Network effects are a winner-take-all proposition; after a technology platform such as Windows, Android, or TCP/IP wins in a market with network effects, it's extremely difficult for others to compete in that market.

Network effects apply to the market for autonomous vehicles (AVs). As the value of the autonomous vehicles marketplace increases exponentially (not linearly), the market grows. The more autonomous cars on the road, the more each one benefits from other autonomous cars on the road (see Chapter 5). For this reason, the returns for companies that win in the autonomous vehicle market will grow with each additional AV in the market.

Network effects also mean that the market can grow at the exponential speed of Facebook, Apple iOS, and Google Android — not at the incremental rates of General Motors and British Petroleum.

### Moore's Law and the Clean Disruption of Energy and Transportation

Electric Vehicles (EVs) are connected, mobile, information technology plat-forms.

The Tesla Model S does over-the-air software downloads to update or patch its operating system.[9] This car has an embedded 3G connection and can also connect via WiFi. In this regard, the Tesla Model S is not very different from your smartphone or tablet computer. Clearly, the Tesla is not your father's Oldsmobile — and clearly your father's Olds manufacturer cannot compete with Tesla.

The electric vehicle is an information technology product. Like many information products, it benefits from Moore's Law (or a version thereof). Moore's Law states roughly that microprocessor technology improves at an annual rate of about 41 percent. According to the law, each subsequent year you can buy a computer that is 41 percent better (faster, smaller, more powerful) for the same dollar amount.

Compound that growth over many years and you get exponentially improving information products such as computers, smartphones, and tablets. This kind of technological improvement rate is the reason microprocessors are a thousand times more powerful than they were twenty years ago and a million times more powerful than they were forty years ago. Exponential technology improvement rates explain why the Silicon Valley has produced industry-busting technologies and companies over the last few decades. You can't compete with exponentially improving products unless, of course, yours is also exponentially improving. If your competitor's rate of improvement is faster than yours, you're toast. It's just a matter of time before the bankruptcy lawyer is knocking at your door. Ask Kodak.

Hendy's Law is the imaging equivalent of Moore's Law. Discovered by Kodak's Barry Hendy in 1998, Hendy's Law states that the number of pixels per dollar doubles every 18 months. This translates to a compounded annual growth rate (CAGR) of 59 percent, which is even faster than Moore's Law. To be competitive in the digital imaging market, you have to match if not surpass this improvement rate.

Apple gets well-deserved credit for innovative and beautifully designed products, but open up one of those iPhones and you'll see exponentially improving technologies to go along with the wonderful design. The iPhone 5S has forty times the CPU performance of the original iPhone[10] for a yearly improvement rate of 85 percent. The iPhone 5S also improved its graphics performance by 56 times for a yearly improvement rate of 96 percent!

Just to keep up with the iPhone, its competitors need to double the graphics performance every year while keeping the same cost!

If your competitor is riding a Moore's Law curve (or its equivalent) and you are not, your competitive offerings are doomed.

It's just a matter of time before the disruption happens to your company. This applies inside as well as outside the industry. Ask Nokia and Blackberry.

Should Detroit auto executives lose sleep if Tesla's electric vehicles are riding a version of Moore's Law? How about several versions of Moore's Law, each corresponding to a different part of Tesla's technology offerings?

To keep up with advances in the electric vehicle, manufacturers of cars with internal combustion engines (ICEs) may pretend to ride a faster exponential curve, but they can't do it. Your father's Oldsmobile may improve at incremental rates (a few percentage points per year), but it can't improve at exponential rates. ICE vehicles are toast. Chapter 4 explores many reasons why the EV disruption wave is clearly coming.

Large, centralized, top-down, supplier-centric energy is on its way out. It is being replaced by modular, distributed, bottom-up, open, knowledge-based, consumer-centric energy. This disruption of the energy industry, coupled with the disruption of the automotive industry, will have a domino effect. Many industries will be disrupted: shipping, trucking, public transportation, car rentals, parking, and insurance. City planning and land management will change dramatically. The ramifications are astonishing.

This is not just happening in Silicon Valley or in digital media. Every significant industry may be disrupted over the next ten to fifteen years.

The century-old energy and transportation industries are on the cusp of disruption. The transition has already started and the disruption will be swift. Conventional energy sources are already obsolete or soon to be obsolete. The business model that enables them cannot compete with the disruptive force of technologies like solar, electric vehicles, and self-driving cars. The innovative business models and participatory culture coming out of Silicon Valley will win the day.

## What about the 100 Years of Oil (or Gas or Coal or Uranium)?

Do you recall the conversation in the 1990s about "peak paper" and whether the United States had enough paper to last a century? Me neither. The web did not disrupt the newspaper industry because we ran out of paper.

Do you recall the "peak vinyl" or "peak CD" crisis? Me neither. The web did not disrupt the music industry because we ran out of either.

The web was just a faster, cleaner, cheaper, more compelling way to produce, store, transmit, and consume content. The newspaper industry and music industry can't compete with the web. The web enabled disruptive products, services, and business models. It created a participatory culture that made the conventional newspaper and music industries obsolete.

The national conversation (if you can call it that) about energy in the media, political circles, and the energy industry is obsessed with whether we are at "peak oil" and whether there is enough natural gas (or nukes or coal) to last for thirty, one hundred, or four hundred years. This conversation misses the point entirely.

The cell phone did not disrupt the old landline telephone industry because we ran out of copper. Enough copper is underground to last one hundred years, but that is not a good reason to invest in landline telephony.[11] Again, the cell phone industry disrupted landline telephony because mobile phones are a faster, cleaner, cheaper, more compelling way to communicate, and they can produce, store, transmit, and consume content.

Just substitute the words oil, natural gas, coal or your favorite conventional energy source for paper, vinyl, or film and you can peer into the future of energy.

The clean disruption of energy and transportation by Silicon Valley's exponentially improving technologies, new business models, and participatory culture is inevitable and it will be swift.

Energy and transportation as we know it today will be history by 2030.

# Chapter 1:
# The Solar Disruption

*"A great many people think they're thinking when they are really rearranging their prejudices."*

*- Aldous Huxley.*

*"First they ignore you, then they laugh at you, then they fight you, then you win."*

*- Mahatma Gandhi.*

*"If you evaluated rooftop solar a year ago, or even three months ago, you are way out of date."*

*- David Crane, CEO, NRG Energy.*

On February 1, 2013, El Paso Electric agreed to purchase power from First Solar's 50 MW Macho Springs project for 5.79 ¢/kWh. "That's less than half the 12.8 ¢/kWh from typical new coal plants," according to models compiled by *Bloomberg*.[12]

Solar costs are going down so fast, solar is already becoming the lowest cost power provider to utilities and to retail commercial and residential consumers. This is true in Australia, the United States, Germany, Spain, and many other markets around the world.

In the United States, new solar capacity has grown from 435 MW in 2009 to 4,751 MW in 2013 for a compound annual growth rate of 82 percent (see Figure 1.1).[13] Solar represented 29 percent of all new generation capacity in 2013, up from 10 percent in 2012 and 4 percent in 2010.[14]

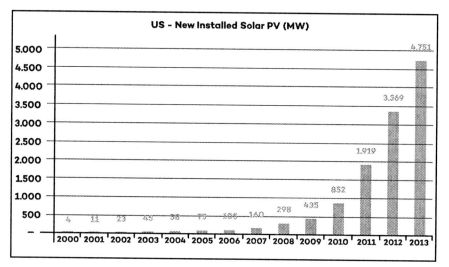

*Figure 1.1—New Installed Solar PV in the U.S. (Data source: SEIA)*

On a clear afternoon on May 25, 2012, solar in Germany generated 22GW, which represented a third of the entire country's power needs.[15] The world record for high penetration of solar was broken on May 25, 2012. The following afternoon, solar generated fully 50 percent of Germany's power, breaking the previous day's world record.

One out of every two electrons flowing through Germany's grid was sunshine just a microsecond before. Astonishing as these penetration numbers were, they have now become commonplace.

In Germany, a country with half the sunshine (solar insolation) as the United States, solar energy drove German wholesale electricity costs *down* by more than 40 percent in 2013 compared with 2008.[16] This represented more than €5 billion ($6.7 billion) in cost savings for the German economy.[17] Solar also pushed volatility down significantly (see Figure 1.2).

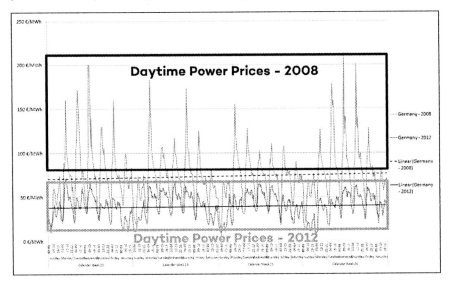

*Figure 1.2—Wholesale power costs in Germany, 2008, 2012, and 2013. Higher solar penetration has lowered both the price of energy and the volatility of energy prices. (Source: Meikle Capital).[18]*

The combination of wind and solar is just as powerful. Shortly before noon on October 3, 2013, solar and wind together provided 59.1 percent of all the electric energy in Germany.[19] Exactly one month later, on November 3, 2013, wind generated more than 100 percent of the power demand in Denmark.[20] You read that right: One hundred percent of the electricity in Denmark was kinetic energy blowing in the wind less than a second before.

Halfway around the world, Australia has set a few world records of its own: one million solar installations in about four years.[21] This represented a market share of about 11 percent of all residential power consumers in the country. It took Germany about twelve years to reach the one million solar installations mark (Germany now has 1.3 to 1.4 million installations[22]). Bangladesh reached its goal of one million solar home systems years ahead of time.

Back in the U.S., Warren Buffett, arguably America's most successful investor, became one of the biggest solar investors in the world. MidAmerican Energy, a subsidiary of Buffett's Berkshire Hathaway, invested $2–2.4 billion to acquire a solar development project that will be the world's largest solar power plant (579 MW) when it opens in 2015.[23]

A relatively small, financially prudent, regional utility that generated most of its power from coal (58 percent) now owns the largest solar power plant development project in the world.[24]

Warren Buffett is usually a step (or two) ahead of Wall Street. Is his investment in solar a leading indicator of the acceptance of solar power by mainstream investors? Warren Buffett isn't waiting around for the answer. MidAmerican also acquired the world's second largest solar power plant project (550 MW) for $2 billion dollars, and a 49-percent stake in a 290 MW solar power plant in Arizona.[25] When MidAmerican went to raise money for these projects, they were all oversubscribed.

At the same time, MidAmerican said they would retire seven coal-fired power plants (see Chapter 10).

Solar has arrived. Solar is on the cusp of disrupting the world's largest industry.

## Cheap, High-Penetration Solar

The success of solar energy has shattered myths and gone a long way to counter the misinformation that the conventional energy industry spokespeople repeat *ad nauseam* — that solar is expensive, that solar is not ready to scale, that solar will take many decades to go mainstream, that solar will bust the grid.

Much has happened in the energy world since I published my book *Solar Trillions* in 2010. Back then it was hard to convince decision-makers that solar would soon be the largest source of energy in the world. Now this is a more acceptable belief. The question is not *if* but *when* it will happen. Even the oil giant Shell now says that solar will be the world's number-one source of energy.[26] Shell's prediction, however, is missing the mark by about seventy years.

We are already in a fast disruptive transition from the century-old model of centralized extraction, resource-based energy production to a clean, distributed, technology-based architecture. This disruption will be all but complete by 2030.

## Exponential Growth of Solar Markets

The installed capacity of solar photovoltaic around the world has grown from 1.4 GW in the year 2000 to 141 GW at the end of 2013 (see Figure 1.3). This translates to a global compound annual growth rate (CAGR) of 43 percent.[27]

In the United States, solar wattage has nearly doubled every year over the past three years. Solar installations in China tripled in 2013.[28]

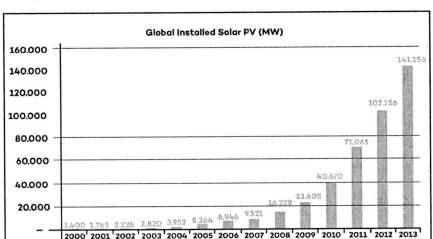

Figure 1.3— *Global Installed Solar PV in MW. (Data source: EPIA and BNEF)*

Germany continues to be the world's leading adopter of solar. As of June 2013, it had 34.1 GW of solar connected to the grid.[29] This figure represents the peak power equivalent of 34 nuclear power plants. The country's maximum monthly power production from conventional sources in 2012 ranged from 50 GW to 65 GW.

By now it has become fairly common for solar to generate between 20 and 35 percent of Germany's total power needs on sunny afternoons. In May 2012 solar generated 20 percent of total power needs 24 days in a row.[30] Germany has added about 7.6 GW since then. Solar power's marginal cost of zero has pushed the cost of wholesale power down.

Europe has continued its transition to clean and distributed energy. In 2011, 48 percent of all new power plant capacity in Europe was solar and 21 percent was wind (see Figure 1.4). Thus in 2011 fully 69 percent of all newly installed power generation capacity in Europe was clean (solar or wind.)

Get used to solar and wind comprising most new power additions to the grid. The Australian Energy Market Operator (AEMO) expects that, by 2020, 97 percent of all new energy generation additions to the grid will be wind or solar.[31] This is the shape of grids to come.

China, the world's largest manufacturer of solar photovoltaic panels, has quickly turned into the world's largest consumer of solar products. After tripling its solar PV demand in 2013, China set a goal of installing 14 GW of solar in 2014. In other words, China is expected to install in one year as much solar as the United States has installed in its history. China's high rate of installation is not uncommon in exponentially growing markets. The United States installed more solar in 2013 than it did in all years prior to the end of 2011.

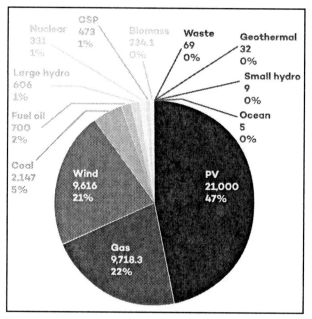

*Figure 1.4 — New power capacity in the European Union in 2011. (Source: European Wind Association)[32]*

The race for solar primacy is on. New Silicon Valley solar companies such as SolarCity, Sungevity, and SunRun are installing tens of thousands of residential and commercial solar plants in California and the rest of the United States, as well as Holland and Australia. SolarCity went public December 2012 (Nasdaq: SCTY) and by August 2013, it had quadrupled its market valuation to about $2.9 billion. To the consternation of conventional energy providers, SolarCity doubled its market valuation again over the next few months.

All these solar installers have developed innovative business models and information technology infrastructures that allow them to scale and grow exponentially. Business model innovation has become a key competitive edge in the solar business.

# The Falling Cost of Solar

John Schaeffer, the founder of Real Goods Solar and the Solar Living Institute in Hopland, California (north of San Francisco), recalls fondly when solar PV was a rarity. At a recent event at the Intersolar Conference, Scheffer told me that PV was rare in the early 1970s, so rare that the U.S. military once knocked on his door asking why he sold panels that used to be the property of NASA. He sold the used solar panels at $90 per Watt. The conventional energy industry would have you believe that solar panels still cost that much.

In 1970, solar photovoltaic panels cost $100 per Watt (see Figure 1.5). Solar entrepreneur Elliot Berman, founder of the Solar Power Corporation, applied a number of manufacturing innovations that dramatically cut the cost to $20/W by 1973.[33] By 2008, solar panels cost $6 per Watt, a 94 percent drop since 1970.

Solar costs have fallen dramatically since then. In 2011 alone, solar PV panel costs went down by 50 percent; in 2012, they fell by more than 20 percent.

By 2013, the market price of solar panels was about 65¢/W. It took just five years for the cost of solar PV to drop by another order of magnitude.

Solar panel costs have dropped by a factor of 154 (from $100/W to 65 ¢//W) since 1970.

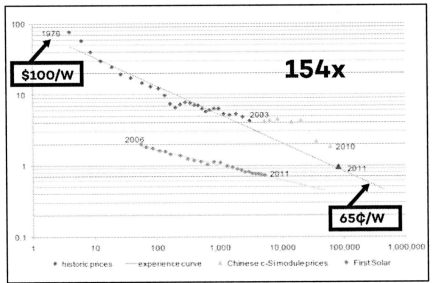

*Figure 1.5—Solar photovoltaic cost curve. Since 1970, PV has improved its cost per watt by a factor of 154 times. (Graph Source: Bloomberg New Energy Finance. Emphasis added by author.)[34]*

The exponential improvement in the cost of solar is something that we have come to expect from the information technology industry, but not the energy industry.

Imagine a world where oil prices followed the same cost curve as solar. In 1970, oil cost $3.18 per barrel;[35] gasoline retailed for $0.36 per gallon.[36] If oil dropped in price at the rate solar has since 1970, a barrel of oil would cost about 2 cents and a gallon of gasoline would cost $0.00234 per gallon. That is, four gallons of gasoline would cost about one cent! In this imaginary world you could fill up your 15-gallon gas tank for 3.5 cents. Instead, oil hovers around $110 per barrel and a tank of gas costs more than $50.

In the real world, solar PV is 154 times cheaper now than it was in 1970 (costs went from  $100/W down to 65 ¢/W), while oil is 35 times more expensive (oil has gone up from  $3.18/barrel to $110/barrel.

Put these numbers together and you find that solar has improved its cost basis by 5,355 times relative to oil since 1970. If you think that any industry can compete with a technology that has improved its cost position relative to yours by more than five thousand times, you're in denial and bankruptcy looms near. This is especially true if your competitor's marginal cost is zero. Just ask Kodak.

Solar has improved its cost position relative to all resource-based energy sources by hundreds or thousands of times since 1970. Table 1.1 illustrates how the solar PV technology improvement cost has fared relative to the cost increase of the world's four main resource-based sources of energy. Later in this book, I examine each conventional source of energy (oil, nuclear, natural gas, and coal).

| Solar Cost Improvement Since 1970 | Magnitude of Improvement | See Chapter |
|---|---|---|
| vs Oil | 5,355 x | 7 |
| vs Nuclear | 1,540 x | 6 |
| vs Natural Gas | 2,275 x | 8 |
| vs Coal | 900 x | 10 |

Table 1.1—Solar PV cost improvement relative to other energy sources since 1970. (See chapter in this book for more.)

Moreover, solar keeps improving its costs basis while the other resource extraction-based energy sources keep increasing their costs.

If you believe that solar doesn't compete with oil, think again. Solar will disrupt oil in two ways (I explain this topic in detail in Chapter 7). The first is by displacing diesel and kerosene, which still provide expensive polluting energy to billions of people around the world. The second is by disrupting the rest of the power industry while the electric vehicle disrupts the gasoline internal combustion engine.

Conventional energy industry insiders hope that solar prices will "stabilize" or even go up.[37] No such luck. Innovation in the solar photovoltaic business is relentless; competition in PV manufacturing is brutal. Even energy giant GE could not take the heat and exited the PV business. It sold its photovoltaic panel technology to industry leader First Solar[38]. GE said it would stay in the solar business but focus on finance and inverters.

The costs of solar have been dropping fast and are expected to continue doing so for the foreseeable future. Costs are going down because of increasing innovation, scale, and competition. The solar experience curve (the solar equivalent of Moore's Law) is about 22 percent. That is, every time industry capacity doubles, costs drop by 22 percent. This rate may have actually accelerated over the last few years (see Figure 1.5).

## Fast Growing Solar Installations

Some energy "pundits" were surprised when Warren Buffett's MidAmerican Energy invested more than $2 billion in solar. They were even more surprised when he kept adding hundreds of megawatts of solar projects. The pundits were not paying attention.

David Crane, CEO of NRG Energy, has certainly been paying attention. "If you evaluated rooftop solar a year ago, or even three months ago, you are way out of date," Crane said. NRG Energy is an $8.8 billion a year company that produces mainly fossil fuel-based electricity to 20 million U.S. households. NRG has invested in solar with gusto.

Crane has been skillfully guiding his company's transition to the solar disruption for a few years. "We believe that in 2014–2016 you'll be able to get power cheaper from the roof of your house than from the grid," said Crane in 2011. "Solar is going to go from this thing that right now is like 0.1 percent of the market to 20 to 30 percent of the overall electricity mix. That's huge."[39]

He added that "solar is as much of a game-changer as I've seen in 25 years in the industry."

MidAmerican Energy is a regional $3 billion a year energy company that uses coal to generate 58 percent of its power. MidAmerican Energy has invested nearly $5 billion in solar and is poised to retire seven coal-burning power plants.

Solar is already past the early stage of the adoption lifecycle in Australia and Germany. In Australia solar has a national penetration of 11 percent; you can find neighborhoods where 90 percent of homes have solar rooftops (see Chapter 3)! In California, solar is already cheaper than grid electricity for a large part of the market. "More than 90 percent of our customers start saving money on day one," reported Danny Kennedy, co-founder of Sungevity, a solar installer based in Oakland.

Furthermore, as soon as 2015, two-thirds of Americans will be able to buy unsubsidized solar power at rates cheaper than what they currently pay for their utilities. Danny Kennedy, who was a guest speaker in my clean energy class at Stanford University, told the class, "By 2016, 40 to 50 million American homes will have to make a decision: Do we want cheap solar power or expensive grid electricity?"

The answer, I think, is clear. Americans want clean and cheap electricity. I have run the numbers and they tell me that, by 2022, there will be 20 million solar installations in the U.S.

The U.S. solar industry is innovating, growing exponentially, and creating a healthy sustainable multi-trillion-dollar industry.

## Why America's Solar Installation Costs Are Higher than Germany's

While solar panel costs have dropped dramatically, the total installed cost of rooftop solar has not dropped nearly as fast in the United States.

Panel costs are so low they are not the main cost factor in residential and commercial solar power plants. More than solar panels, so-called soft costs now account for a higher percentage of total installation costs. Soft costs include the cost of permits, regulations, taxes, interconnection fees, inspections, and installation.

Permits and regulations, for example, add a relatively significant slice to the cost of installing solar. A study by solar installer SunRun found that the cost of getting a permit added $2,516 ($0.50 per Watt) on average to a solar installation.[40] That's almost as expensive as the panels themselves. As I write this, solar panels are about $0.65 per Watt and are projected to drop to $0.36 per Watt by 2017.[41]

In Germany the total costs of a small (<10kW) solar PV installation dropped to $2.26 (1.698 Euro) per Watt in late 2012.[42] In Australia the average cost of installed PV for a 5kW residential system was even lower, at A$1.76/W (US$ 1.62/W) in July 2013.[43]

In Perth, installed solar was as low as A$1.38 (US$1.27/W).

The total cost of a residential or commercial solar installation (per Watt installed) in the United States has consistently been much higher than Germany. Americans pay about twice as much for the equivalent power plant in Germany.

Given the fact that panels, inverters, and most of the hardware are globally traded and similarly priced in both markets, the U.S. buyer is paying $2.8 more per Watt in capital costs than her German equivalent, according to a report by the Lawrence Berkeley National Laboratory.[44]

Even if solar panels cost nothing, American consumers would still pay more in "soft costs" than Germans or Australians pay for the whole solar system. In fact, American "soft costs" are nearly twice the cost of the whole solar system in Perth, Australia.[45]

Despite the fact that America's solar installation costs are $2.8 per Watt more expensive than equivalent global markets, the solar PV market in the U.S. has nearly quadrupled over the last few years. This bodes well for the future of clean energy. Just by matching Germany or Australia, U.S. solar costs will drop at least another 50 to 60 percent. The U.S. Department of Energy's Sunshot Vision program foresees $1.5/Watt by 2020.[46] Basically the DOE hopes the U.S. will take six years to catch up with today's solar prices in Australia.

If historical costs and learning curves are any indication, solar costs will drop further and faster than market prognosticators anticipate.

## Unsubsidized Solar vs. Subsidized Utility Rates

The world's largest solar panel vendor, First Solar, said its global strategy is to build solar in markets without government subsidies.[47] Unsubsidized solar is already cheaper than subsidized fossil fuels and nuclear power in hundreds of markets around the world. That's an interesting development.

The total capital cost of building a utility scale solar power plant has dropped below $2/Watt and is getting closer to the $1/Watt mark. The decision about switching to solar energy will not be about being green but about saving green (dollars). The residential market alone may be worth a trillion dollars. The U.S. commercial solar market may be larger. Similar trillion-dollar disruptive market opportunities await around the world.

Greentech Media expects solar panels to drop to 36 cents per Watt by 2017.[48]
Citibank expects panels to drop to 25 cents/Watt by 2020 (see Figure 1.6)[49]

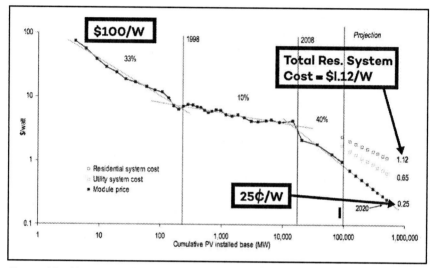

Figure 1.6—Historic and projected cost of solar photovoltaic. (Source: Citigroup,
emphasis added by author)[50]

Citigroup further predicts that the total installed cost of solar PV will be
65 cents/Watt for utility scale solar plants, while the total installed cost of
residential systems will be $1.12/Watt.

## Lancaster, a Case Study in the Future of Energy

What does inexpensive, unsubsidized solar mean for the energy industry?
For a preview of the disruption that will occur, consider the city of Lancaster,
California. On March 26, 2013, the city council of Lancaster approved a change
in the zoning laws by which every new home must be built with solar panels.

Located a hundred miles northwest of Los Angeles, Lancaster is an ambitious
city. In early 2012, Jason Caudle, the city's deputy manager, told me that his
city of 155,000 residents had built 23 MW of solar and approved an additional
100 MW that were waiting for interconnection permits. Assuming all those
permits are granted, Lancaster will have 794 solar Watts per resident, or 60
percent more than Sonoma County, the state leader. Extrapolate that number
to the rest of California and the Golden State would generate 30.5 GW of solar.
California's peak energy demand in 2011 was about 60 GW. By generating
30.5 GW, more than half of California's peak power could come from solar.[51]

But Lancaster is more ambitious than that. Lancaster wants to be the world's first net-zero emissions city. To do that, Lancaster effectively needs a solar capacity of nearly 600 MW. Do the math and you see that this city of 155,000 residents wants to build about 4 kW of solar per resident.

To reach its net-zero goal, Lancaster would have to multiply its state-leading numbers by 300 percent. To meet that goal, the city changed its regulatory process. A standard residential solar installation would be approved over-the-counter in just fifteen minutes; the fee would amount to $61 (permit = $31, travel and documentation = $22, and issuance = $6).

Jason Caudle, Lancaster's deputy manager, told me that Lancaster planned to build 50 MW of solar and sell its output to neighbors at 8.5 ¢/kWh. Is that possible? Following is my calculation of the levelized cost of generating solar electricity (LCOE) in Lancaster. Since the cost of capital (COC) is the main determinant of solar costs, I plotted the LCOE vs COC. Here are my assumptions for 2020:

- Solar insolation: 2400 kWh/m²/year
- Solar panel efficiency: 15.9% (same as 2013)[52]
- Installed cost per PV Watt: $1.12[53]
- O&M: 1% of plant installed (turnkey) cost
- Insurance: 0.3% of plant installed (turnkey) cost

In 2020, the cost of rooftop solar generation in Lancaster will range from 3.47 ¢/kWh to 6.62 ¢/kWh, depending on the cost of capital (see Figure 1.7). No subsidies apply.

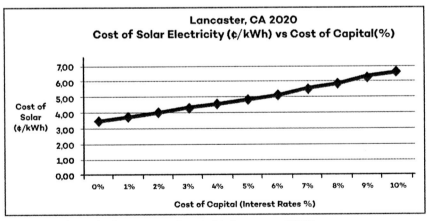

Figure 1.7—Cost of solar vs. cost of capital in Lancaster, 2020.

The only assumption in my calculation that differs significantly from today's U.S. market prices is the installed cost per PV Watt. My calculation uses Citi's projection of $1.12/W. How optimistic is this number?

The installed cost of residential solar in Perth, Australia in August 2013 was already as low as A$1.38 (US$1.27/W).[54]
Basically, Citi projects that the U.S. needs six years to improve slightly on Australia's 2013 costs.

In Lancaster, solar panels will be an integral part of every new house, like windows and doors, so they can be financed along with the mortgage on the house where the panels are installed. Assuming mortgage rates in the U.S. are 3.47 percent (15 years fixed, 0 points), the cost of residential rooftop solar in Lancaster is about 4.4 ¢/kWh. No form of energy generation can deliver energy to a home at this low price. The cost of transmission and distribution alone is higher than 4.4 ¢/kWh. Utilities won't have any business when retail solar generation reaches 4.4 ¢/kWh.

Extraction-based centralized power plants (nuclear, gas, coal, oil) will certainly not have a retail business at that point. As is, they are already being clobbered by solar in the wholesale markets. Conventional power plants will go by a different name in the investment industry: stranded assets.

Can't believe that the retail cost of solar will be less than 5 ¢/kWh by 2020? Lancaster is already paying 10 ¢/kWh for the solar generated on its buildings, according to Lancaster's deputy manager, Jason Caudle.[55] "The utility companies were charging us 18 ¢/kWh," he said, "so we're already saving money."

Solar photovoltaic costs are going down, installation costs are going down, the cost of capital for solar is going down, and maintenance costs are going down. It would be surprising if Lancaster didn't get to 5 ¢/kWh sooner than 2020!

Importantly, the leadership in Lancaster is ambitious and fully behind solar. "We will be a net-zero city within three years," said Mayor Rex Parish.  To accomplish this goal, the city has built a solar pipeline of 700 MW.[56]

If it achieves its goal, Lancaster will not just be a net -zero city; it will be a net exporter of energy. It will also have achieved its solar energy goal in less than a decade.

## How Quickly Can Solar Disruption Happen?

Is the Lancaster experience possible on a national or global basis? If the U.S. had Lancaster's solar adoption of 4kW per resident, it would have 1.2 TW of solar. A hundred percent of the country's peak power would be solar. Lancaster will likely reach its goal before 2020. How about the rest of the nation?

Skeptics argue that it takes many decades to build a solar infrastructure. Even if rooftop solar were far cheaper than the retail grid electricity that the local utility provides, you can't build tens of millions of solar rooftops in a decade, skeptics argue. What the skeptics don't understand is, when disruption happens, it happens swiftly. Just ask anyone at your favorite camera film, telegraph, or typewriter company.

I have explained that Australia provides an example of how quickly a market can adopt solar. Australians went from hardly any residential solar to one million solar homes in about four years. About 2.6 million Australians, 11 percent of the population, now have rooftop solar.[57]

The population of the United States is about ten times the size of Australia's population, so an equivalent penetration rate in America would result in about 10 million solar installations in four years. Do the energy "experts" believe that Australians can do it but not Americans?

There are about 300,000 solar installations in the U.S. Can the solar industry realistically grow to 10 million or 20 million installations as quickly as Australia? Do we have the supply chain and the labor to install 10 to 20 million rooftop solar systems in such a short time?

For comparison purposes, to find out whether the U.S. can adopt solar at a fast pace, consider the experience of another disruptive rooftop-oriented technology industry: satellite television.

## DirecTV's Rooftop Disruption

In 1994 a new company called DirecTV launched a multi-channel television programming service that was beamed directly to homes via rooftop satellite dishes. The company's direct broadcast satellite (DBS) service provided homes with an alternative to landline cable television programming.

Satellite dish technology had been improving exponentially for years. In 1984 "home dishes" were up to 10 feet in diameter, cost $5000, and could transmit up to 27 channels. By 1994 they had shrunk to just 18 inches, cost $700, and could transmit up to 175 channels.[58]

Could DirecTV's rooftop service compete with landline-based monopoly cable TV providers? If so, how fast could sales grow? After all, workers would have to climb on the roofs of houses, make sure the roofs were stable, lay out cables, and so on. Would neighbors complain about these strange contraptions sprouting on the rooftops?

DirecTV's brand-new, unlikely technology winner went from market launch to nearly 10 million customers in the first five years (see Table 1.2). Its market penetration reached just over 10 percent of homes with television sets.

| Year | Total Number of TV Homes Acquiring DBS (millions) | Market Penetration % of TV Homes with DBS (%) |
|---|---|---|
| 1 | 1.15 | 1.21 |
| 2 | 3.076 | 3.21 |
| 3 | 5.076 | 5.25 |
| 4 | 7.358 | 7.55 |
| 5 | 9.989 | 10.16 |

Table 1.2—Rooftop direct-broadcast-to-satellite (DBS) adoption. (Source: Frank M. Bas et al. )[59]

A solar photovoltaic installation is more complex than a satellite dish installation, but the fact is a trained crew can complete a home rooftop solar installation in a few hours. Notice that both the timeframe (five years) and the market penetration (10 percent) are close to the Australian residential solar PV adoption numbers (four years and 11 percent).

Using the logistics technology that DirecTV had in 1994, ten million solar rooftops could be built in the U.S. in just five years.

Computers are literally a thousand times more powerful now than they were in 1994. (That's Moore's Law at work.) The logistics industry has improved by orders of magnitude since DirecTV launched its (DBS) service over two decades ago. DirecTV didn't have mobile Internet communications, a cloud infrastructure, and big data analytics. The concept of Enterprise Resource Planning (ERP) was barely getting started. Google Maps and Google Earth were years away from being conceived.

Furthermore, best-of-breed logistics technologies are now available on the cloud as a service. There are now dozens of solar companies with the talent and funding to scale their operations.

If DirecTV, a single company, could install 10 million satellite rooftops in five years using what is now ancient (1994) logistics technology, there is no technical reason why a single company can't install 10 million or 100 million solar PV rooftops in five or ten years using 2014 logistics and mapping technologies. Even if a single company can't do it, the top ten solar installers as a group certainly can. The ten companies could include solar installation/third-party finance companies such as SolarCity, Sungevity, SunRun, Vivint Solar, and Verengo. As well, perhaps some of the same Australian companies that built the million solar rooftops there in just four years could lend a hand.

When the cost of solar hits the point of no return, the impediments to building 10 million, 40 million, or 100 million solar rooftops in less than a decade will certainly not be technical. The impediments will be legal, political, and regulatory (they will likely be placed there by the incumbent energy companies).

## The World's First Around-the-Clock Solar Power Plant

In the near future solar power plants will be as plentiful as personal computers and cell phones are today. They will generate energy on demand. In June 2011, I went to Southern Spain to witness Gemasolar, a solar power plant capable of generating solar electricity around the clock.

Built by Torresol Energy, Gemasolar is a 19.9 MW solar plant with fifteen hours of thermal energy storage that generates electricity anytime during the day or night — at 11 p.m., 1 a.m., or 3 a.m. (You can watch a YouTube video of my visit to Gemasolar at http://www.youtube.com/watch?v=GhV2LT8KVgA.)

Gemasolar's expected production is 110,000 MWh per year—enough energy to fully power 25,000 households, according to Santiago Arias, Torresol's co-founder and chief infrastructure officer. Because it can store energy, this 19.9 MW is equivalent to a 50 MW solar power plant without storage, according to Arias.

### Solar Salt Batteries

Gemasolar's battery consists of two tanks of molten salt thermal energy storage. The battery allows the solar plant to generate on-demand electricity — at night, when it's cloudy, when it rains, days or weeks later. Molten salt energy storage (MSES), also called "solar salt" batteries, are thermal, not chemistry-based batteries like the Lithium-ion battery that powers electric vehicles such as Tesla's Model S (see Chapter 4).

MSES uses a combination 60-percent potassium nitrate and 40-percent sodium nitrate that retains 99 percent of the heat for up to 24 hours. Another way to put this number: This battery loses just 1 percent of the heat energy per day.

Potassium nitrate happens to be environmentally safer and cheaper than most chemical-based battery alternatives. In the European Middle Ages, potassium was used to preserve food. Potassium nitrate is still used in the production of corned beef[60], in toothpaste (for sensitive teeth), and in garden fertilizers. MSES capital costs are relatively low, clocking in at about $50 per kWh, a tenth of the cost of a Li-on battery.

Gemasolar is not the world's first commercial solar plant to use MSES. If you drive another 300 Km (186 miles) southeast from Gemasolar on Andalucia's A94 highway, you come to Andasol-1, a 50 MW CSP plant that has been operating with a 7.5-hour battery since July 2009. Gemasolar basically doubled the battery availability to 15 hours.

Santiago Arias expects Gemasolar to produce electricity at about 6,400 hours per year with a capacity factor of 75 percent. By comparison, the Hoover Dam has a capacity factor of about 23 percent, and China's massive Three Gorges hydro-electric power plant has a capacity factor of about 50 percent.[61]

According to a 2003 study by Clemson University Professor Michael Maloney, in 2003 the capacity factor of nuclear reactors in Japan, France, and the U.S. was in the 65 to 72 percent range; the load factor worldwide was 69.4 percent.[62] These numbers were calculated before the nuclear disaster at Fukushima Dai'ichi brought the whole Japanese nuclear industry to its knees.

## How Energy Storage Changes Everything

Santiago Arias, Torresol's chief infrastructure officer, started building power plants 38 years ago. On the subject of the electricity market, he gets excited about the impact of a solar power plant that can operate around the clock. "The maximum demand for electricity takes place during the evening on the hottest days of the year," he told me in his office at the power plant. The market pays a premium price for electricity during those peak hours. A solar power plant generates the most energy precisely when the most energy is needed during hot sunny days.

Arias said, "The ability to store energy when the sun it at its peak and deliver it when the market demand is at its peak changes everything in the power market. My fuel cost is zero. Natural gas can simply not compete with us."

## Solar at Night: Baseload (24-7) Solar is Here

The United States is building a massive solar infrastructure of both concentrating solar power (CSP) and solar photovoltaic panels. Not including residential or commercial installations, the U.S. has more than 4,200 MW of projects under construction and more than 23,000 MW projects under development, according to the Solar Energy Industries Association.[63]

As of early 2014, Los Angeles-based SolarReserve is planning to commission the world's largest baseload solar power plant (110MW) in Nevada. Built in the town of Tonopah, this solar plant is about five times larger than Spain's Gemasolar. Tonopah Solar will sell its energy to power Las Vegas into the evening hours. The Las Vegas Strip will soon be lit with solar energy.

On a much smaller scale, the South Pacific island nation of Tokelau became the first country in the world to go 100 percent solar. Spread over three atolls, Tokelau's solar PV is stored in battery banks for use at night. Tokelau made the transition from 100 percent powered by diesel to 100 percent powered by solar in less than one year.

## The Solar Disruption Is Here

Much of the rest of this book explores the many reasons why solar will disrupt the energy industry. I have been here before. The exponential growth of solar reminds me of my days at a decidedly smaller Cisco Systems.

I worked in business development at Cisco Systems in 1993. I still recall looking at the growth of the Internet and thinking, "If this market continues to grow at this rate, it's going to be on a billion devices within a decade." Back then many people looked at me strangely when I said that. Most people did not even know what the Internet was, let alone believe that they themselves and a billion other people would soon own an Internet-connected device.

In 1990 there was one (yes one) web server and one web browser. They both ran on Tim Berners-Lee's NeXT Computer at CERN in Switzerland.[64] The first web server outside Europe saw the light of day December 1991 at Stanford Linear Accelerator Center in Palo Alto, California.[65] Today there are more than ten billion interconnected devices.[66]

Solar is on a similar exponential trajectory. Globally, solar PV installed capacity has grown from just 1.4 GW in 2000 to 141 GW in 2013. This represents a compounded annual growth rate (CAGR) of 43 percent.

Should solar PV continue to grow at a 43-percent annual clip, the solar capacity installed base will be 56.7 TW by 2030. This is approximately the equivalent of 18.9 TW of conventional baseload power. World demand for energy is expected to be 16.9 TW by 2030, according to the US Energy Information Agency.

Should solar continue on its exponential trajectory, the energy infrastructure will be 100-percent solar by 2030.

This statement, that the world will be 100-percent solar, usually generates the same befuddled looks I used to get when I forecasted a billion Internet nodes twenty years ago. The question is: Can solar keep growing at this exponential rate for another ten or twenty years?

The answer is that the solar growth rate could actually accelerate. As a rule, when a technology product achieves critical mass (the point of no return), its market growth actually accelerates.

As solar approaches critical mass in many markets around the world, a virtuous cycle of market adoption will result that makes solar growth accelerate:
- Capital availability increases and cost of capital decreases
- Local, distributed energy generation increases
- The architecture of energy flips from centralized to distributed
- Enabling technologies such as sensors, artificial intelligence, big data, and mobile communications improve exponentially
- The cost of resource-based energy increases
- Complementary markets such as wind, electric vehicles, and self-driving cars increase exponentially
- Investments increase in other storage technologies shared by solar, wind, electric vehicles, and self-driving cars
- The architecture of energy becomes more and more distributed
- The conventional command-and-control energy business model enters a vicious cycle of rising prices and stranded assets

The trends leading to this all-solar (and wind) scenario are already playing out. According to a report by the Australian Energy Market Operator (AEMO), 97 percent of all new energy generation additions to the grid in 2020 will come from wind or solar power.[67] After that it's all about retiring old dirty plants until the whole grid becomes wind and solar.

The disruption caused by exponentially improving technologies, new business models, and a participatory finance and energy culture is coming. Many conventional "energy experts" will tell you otherwise. Or they will tell you that the disruption will be expensive or take many decades or a century. The same "telecom" experts used to make similar predictions.

In 1985 AT&T hired McKinsey & Co, a leading management-consulting firm, to predict what the cell phone market in the United States would be like in the year 2000. McKinsey predicted that fewer than one million cell phones would be sold that year.

AT&T, which practically invented the field of wireless communications, made the decision not to enter the business based on what they thought was a tiny market opportunity. The "smart" telecom money was against cell phones in the mid-1980s. In 2000 the actual number of cell phones was 106 million. The "smart telecom experts" predictions were off by a factor of a hundred. AT&T eventually paid $12.8 billion to buy McCaw and re-enter the field in 1994.

More than six billion people own cell phones in a world of 7 billion people.[68] To put this in context, consider that only 4.5 billion people globally have access to a toilet, according to the United Nations. So more people have access to a cell phone than a toilet. More than 2,800 years have passed since the invention of the toilet in Minoa; more than two thousand years have passed since Rome built a working sewer system.[69] Yet digital cell phone use surpassed toilet use in two decades.

You have to admit that building a distributed, over-the-air, bit-based infrastructure is easier than a centralized, pipeline-based, atom-based one.

There are many examples of products that disrupted trillion-dollar industries in two decades or even two years. Kodak was still a photography giant in 2003. The smartphone market was a niche market before Apple released the iPhone in 2007. Tablet computing barely existed before Apple came out with iPads in the year 2010.

The iPhone and iPad were characterized by exponential growth. Exponential growth is not intuitive; it's hard to imagine. But when a product category grows at an exponential rate you better pay attention. Ten or twenty years of exponential growth can have a huge disruptive impact.

Solar will disrupt energy. But market disruptions are not just about disruptive technology innovations blowing away existing products and industries.

Business model innovation is as important as technological innovation when it comes to disrupting an incumbent's businesses. To understand business model disruption, you need to understand finance and financial innovations.

# Chapter 2:
# Finance and the Disruption of Energy

*"Business model innovation is more important than technology innovation."*

*- Abe Reichental, CEO 3D Systems (2013).*

*"The most common way people give up power is by thinking they don't have any."*

*- Alice Walker.*

*"The truth is no online database will replace your daily newspaper."*

*- Clifford Stoll, astronomer and author.*

In 1918 one in thirteen American families owned a car. Eleven years later, 80 percent of American families owned one. The main reason the U.S. auto market went from early adopter to nearly full penetration in slightly over a decade was an innovation launched by General Motors. This innovation had nothing to do with advances in engine technology, new transmissions, or other technological innovations.

In 1919, GM partnered with DuPont to form the General Motors Acceptance Corporation (GMAC). The purpose of this corporation was to offer car loans to car buyers.[70] Seven years later, 75 percent of all car buyers bought cars on credit.

It was a financial innovation on the part of GM and DuPont, not a technological innovation, that made cars affordable to the mainstream American buyer. In other words, the transportation industry was disrupted by a business model that had never been seen before. The energy industry, likewise, is being distrupted by a new business model.

## New Business Models for Solar

In 2008, a company called SunEdison introduced the concept of solar-as-a-service. Residential and commercial solar power buyers would no longer need to invest their capital to purchase solar panels.

SunEdison offered to finance, install, own, and maintain the solar panels on its customers' rooftops. The company offered to do this with no money down. Homeowners did not have to take any technological, financial, or maintenance risks. They signed a 20-year contract with SunEdison, and at the end of 20 years, customers had the choice of purchasing the equipment at a deep discount or having the equipment taken off the roof.

Soon after SunEdison, another Silicon Valley installer called SolarCity created the SolarLease, and the solar market exploded. Under the SolarLease plan, instead of buying the equipment, customers leased it. The concept caught on and Silicon Valley companies such as Sungevity and SunRun joined SunEdison and SolarCity in offering "solar leases" or solar power purchase agreements (called solar PPAs in the industry).

These solar companies offered consumers a financial innovation: solar-power-as-a-service agreements with zero-money-down and flat payments for the duration of the contracts (14 to 20 years).

There were two main types of solar-as-a-service contracts:
- The solar PPA (solar power purchase agreement). The customer agrees to buy electricity generated by the solar panels at a fixed rate per kWh for the duration of the agreement.
- The solar lease. The customer agrees to pay a fixed monthly lease for the panels regardless of power production.

Under both contracts, the solar company would purchase, finance, install, and maintain the panels and complementary technology at no cost to the consumer. Solar consumers now had the same options that car buyers have had for almost a century: they could purchase solar power with cash, buy it on credit, or lease it.

These financial innovations worked. The solar PV market in America nearly doubled each year after 2009 for a compounded annual growth rate (CAGR) of 97 percent. Most of this growth came about due to third-party solar finance companies.

By mid-2012, the percentage of residential solar installations that were financed by third parties went from single digits to 75–80 percent in California, Colorado, and Arizona (see Figure 2.1). Those percentages are still increasing. About 80 percent of residential and commercial installations are now financed by third-party companies.[71] In Colorado, the number is closer to 90 percent.

This business model innovation was so successful, in 2009 Scientific American named the "no money down solar plan" one of its "20 World Changing Ideas."[72]

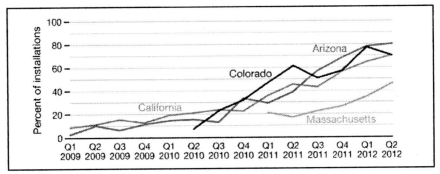

Figure 2.1—Percentage of third-party financed residential solar photovoltaic installations. (Source: SEIA and GTM Research)

Third-party finance also saved the consumer from having to go to the bank and ask for credit to finance a solar installation.

Moreover, third-party finance helped change the conversation about solar power. For years, energy companies used the concept of "payback" to

discourage customers from installing solar panels. They told customers it takes many years for solar energy to pay back the initial capital investment. This concept of payback was based on two assumptions:
- The customer would pay cash (or borrow) to purchase the solar installation.
- The customer would pay more for solar electricity initially, but as conventional power prices went up and solar costs stayed flat, the customer would eventually save enough money to recover his or her initial cash investment.

Third-party finance turned this concept on its head. "What does 'payback' mean when you don't have to invest anything and you start saving money on day one?" asked Danny Kennedy, co-founder of Sungevity, speaking to my class at Stanford University in 2012.

By the time it went public in December 2012, SolarCity [Nasdaq: SCTY], an early leader in third-party solar financing, had raised more than $1.3 billion in project finance capital. In June 2013, to finance residential solar, SunRun raised $630 million from JPMorgan Chase, US Bank, and others.[73] Vivint Solar has raised more than $740 million. Clean Power Finance manages $500 million to finance its solar installations.[74]

SolarCity went public in December 2012 at $9.25 per share. By August 2013, when it announced that installed solar watts had grown by more than 144 percent and lease revenues had gone up 78.8 percent[75], stock in the company quadrupled to about $40 to give the company an implied valuation of $2.9 billion.[76] The stock doubled again over the following few months.

Just like the auto industry did a century before, solar literally went through the roof due to a business model innovation.


## The Drop in the Cost of Solar Panels

The similarities between the early auto industry and today's solar industry don't end there. Besides auto loans, another key reason the auto market went through hyper-growth had to do with lowering costs. Costs went down significantly due to competition, scale, and innovation.

In 1908, a Ford Model T retailed for $850, which put it out of reach for median wage earners in America. By 1914, however, the cost of a Model T dropped to $490. By 1921, the car cost $310 — a drop of 62 percent in thirteen years.

Compare the drop in auto prices from between 1908 and 1921 to the drop in solar installations. In 2009, a "typical" 4kW residential rooftop solar installation cost about $32,000 (before incentives).[77] At this price, the cost of installing solar panels was higher than the median income of wage earners in America, which in 2009 was about $26,684.[78] However, solar panel costs dropped by 50 percent in 2011 alone and have dropped by about 90 percent since 2008. Solar costs have dropped further and faster than the cost of buying a car did in the previous century.

Once it hit critical mass, it took a little more than a decade (1908–1921) for the auto industry to completely disrupt the horse-based transportation business. We should expect the same adoption curve with solar energy. What's holding it back?

## Capital Costs and the Costs of Capital

While "soft costs" are the main portion of the capital costs of solar, the cost of financing a solar plant has become the most important factor in the cost of solar electricity (the levelized cost of electricity, or LCOE.) When you purchase a house by taking out a 30-year mortgage, most of your monthly payments go to cover interest. The same is true of financing a solar power plant.

The cost of fuel (the sun) is zero. The ongoing costs of operations and maintenance (O&M) are close to zero. You do need to wash the panels once in a while (the rain can also take care of that) and replace the inverter every ten years or so. (The inverter takes the panel DC power and turns it into AC power, which is what is used in homes.) Consequently, the cost of borrowing the capital (the interest rate) constitutes most of what a solar user pays for solar energy.

According to Lyndon Rive, CEO of SolarCity, one percentage point in the cost of capital reduces project costs by the equivalent of 20 cents per Watt. What's holding solar back? Basically, Rive believes that the industry has been paying "credit card interest rates" to finance solar installations.

Still, financial innovations have proven that they can quickly unlock the solar market. The concepts of solar PPA and solar leases that I mentioned earlier are but the beginning of a menu of options that are becoming available to solar developers.

# Case Study: PACE Funding in Sonoma County

One of the first programs to prove the importance of cheap and accessible capital for clean energy production was led by Sonoma County, north of San Francisco.

At the end of 2011, Sonoma County had 500 solar watts per resident and 4.5 solar installations per 100 residents.[79] These numbers don't sound like much, but if you extrapolate Sonoma County's numbers to California's 38 million residents, the state would have 19 GW and 1.7 million solar installations. The actual numbers in California were less than a tenth of those numbers. In fact, Sonoma County's numbers surpassed the Golden State's 2020 distributed generation goal (12 GW) by 45 percent and the number of solar installations (one millions solar roofs) by 70 percent.

Sonoma achieved these numbers in less than three years, in the midst of a national financial crisis, despite fierce opposition from the U.S. Federal Housing Finance Agency. Sonoma County was clearly doing something right.

The centerpiece of Sonoma's clean energy program was the Sonoma County Energy Independence Program, or SCEIP. SCEIP is a PACE (property assessed clean energy) program. It was established in March 2009. Its goal is to improve "performance in 80 percent of Sonoma County homes and commercial spaces to the highest cost-effective efficiency levels."

Conceptually, PACE is a local municipality finance program by which municipal governments can tap private capital markets in order to finance energy efficiency and clean energy projects for homes and commercial properties. Funding is done through an assessment of property taxes on the homes and commercial properties.

PACE is a variant of the "no money down solar" that Scientific American called one of its "20 World Changing Ideas." PACE is an accessible, low-interest rate financing mechanism that makes it easy for homeowners and business owners to improve their buildings with clean and efficient energy.

Sonoma County's PACE program had the following characteristics:
- The financing takes the form of an assessment, not a loan. Unlike a loan, an assessment is attached to the property rather than an individual. This dramatically reduces the financing risk.
- The assessment takes the form of a lien, so the payback responsibility automatically transfers to subsequent owners of the property if the property is sold before the assessment is fully paid off.

- Financing must be 10 or 20 years and is paid through an assessment on the owner's annual property taxes.
- Improvements must be permanently fixed to the property.

PACE financing was originally designed to get around the fact that investments in energy efficiency and clean energy had longer-term payoffs while the capital costs generally had to be borne up front. PACE financing was created in 2005 in Berkeley, California and soon spread to 23 states around the nation.[80]

The Sonoma County Energy Independence Program (SCEIP) was created with $60 million in funding; $45 million came from the county treasury and $15 million from the county's water agency. SCEIP funded $58.5 million worth of energy retrofit projects, 2,855 residential and 87 commercial. The county estimated that 79 percent of the 682 jobs that SCEIP generated were local jobs.

My focus is on solar, but SCEIP funded more than a thousand non-solar projects as well, including more than 500 window and door installation projects, 200 HVAC (heating, ventilation, and air conditioning) projects, and 200 sealing and insulation projects.

Sonoma County proved that PACE financing works. It worked even in the middle of a national recession and financial crisis. SCEIP was not funded with taxpayer money but with bonds raised from private investors. SCEIP actually made a small spread from financing projects.

Other states and municipalities noticed the success that Sonoma County had with PACE funding. PACE is a municipal government program and, as such, it needs legislation at the state level to keep it going. At one time PACE funding was active in 23 states and was being considered in 20 more states.[81]

What would a PACE program look like nationally? If the entire United States achieved Sonoma's average of 500 solar watts per resident and 4.5 solar installations per 100 residents, the United States would have 159 GW of solar and 14 million solar installations. This wattage is equivalent to the peak power of 159 nuclear power plants — roughly all the nuclear plants in the U.S. and Japan put together.

Furthermore, the U.S. could achieve this production in three years like Sonoma County did. This is a truly disruptive scenario. Nuclear would be redundant and out of business in three years. The power utilities can't compete with this.

The Federal government wants the U.S. to build a clean energy economy — a "sun shot" to match John F. Kennedy's "moon shot."

Sonoma County's PACE quietly showed it can be done. Sonoma provided the recipe. Will the Federal government take the lead?

Oddly enough, it was a Federal government agency that brought the quick adoption of PACE programs around the nation to an abrupt halt. The Federal Housing Financing Agency (FHFA), which oversees the quasi-government mortgage financing agencies Fannie Mae, Freddie Mac, and the Federal Home Loan Banks, objected to PACE. The FHFA argued that "first liens established by PACE loans … pose unusual and difficult risk management challenges for lenders, servicers and mortgage securities investors."[82]

I asked Diane Lesko, the SCEIP program manager, about FHFA's claim that PACE-funded energy retrofits are high-risk investments. She said that the default rate on the Sonoma County PACE program was 1.1 percent, whereas the mortgage default rate at the time was near 10 percent.

Solar was a better investment risk than home mortgages. Yet Fannie Mae and Freddie Mac — the taxpayer-backed financial services that made hundreds of billions of dollars of Mickey Mouse mortgages possible and nearly brought the entire U.S. economy to its knees — told American banks not to finance an asset with a 1.1-percent default rate.

PACE succeeded as a financial innovation, but politics got in the way. The Federal Government snatched defeat from the jaws of victory.

I asked SCEIP's Diane Lesko what was the most important ingredient in building an energy-independence program into a winning program. "Political will," she answered without missing a beat. "You need leadership coming together to achieve our common goals."

Despite the opposition of the FHFA to residential PACE loans, many municipalities around the country adopted PACE programs for commercial (not residential) solar. Miami-Dade County in Florida, for example, announced a $550 million commercial PACE program to be led by Ygrene Energy, a Santa Rosa (Sonoma County)-based financial services company.[83] Ygrene also targeted a $100 million of PACE funds for Sacramento, California.

Sonoma County proved that accessible, low-capital-cost financing could unlock and propel clean energy markets. What solar needed was a financial innovation of the sort that disrupted the auto industry a century ago, and Sonoma County provided it. Sonoma County also showed that the disruption could happen quickly.

# Participatory Finance: Crowd Funding for Solar

Financial innovations like PACE, the solar PPA, and the solar lease showed how to create fast growth in the residential and commercial-scale solar markets. Meanwhile, the web-enabled sharing economy has created other instruments that can be used to finance solar installations.

On December 10, 2012, a San Francisco non-profit organization called Re-Volv launched a fundraising campaign through a crowd-funding website called Indiegogo.com (also located in SF). Re-Volv wanted to raise $10,000 in six weeks to help finance a solar rooftop installation for a community organization. Re-volv raised the $10,000 from a hundred donors in just three weeks. What's more, Re-Volv exceeded its fundraising goal by 50 percent for a total of $15,391 raised.[84]

What did donors get for their money besides the satisfaction of helping a community organization? The two people who contributed $1,000 got a tour of Re-volv's office in San Francisco and a chance to meet the Re-Volv staff. Those who contributed $500 got a phone call from Executive Director Andreas Karelas. If you contributed $50 you got to choose between a copy of my book Solar Trillions and Paul Wapner's Living Through the End of Nature.

Most of the people who donated to this crowd-funding campaign did so to help make the world a better place. In fact, of the 75 people who donated $50 and were eligible to claim a book, only 24 did so.

Three months after Re-Volv's crowd-campaign ended, I got an email from Andreas Karelas telling me that his company had agreed to finance a 10 kW solar rooftop installation at the Shawl-Anderson Dance Center in Berkeley. The installation would cover 100 percent of the dance collective's electricity needs for the duration of the contract. In turn, Re-Volv would use the proceeds from the lease to finance other solar installations (thus the name Re-Volv).

A year later, Re-Volv kicked off its second fundraising campaign, this time for $55,000. When the final results came in, 303 contributors from twenty American states and seven countries helped Re-Volv surpass its goal by more than $1,000. The money was used to install a 22 kW solar system for the Kehilla Community Synagogue in Oakland. "We will save them over $130,000 in electric bills over the life of the system," said a happy Andreas Karelas.

Re-Volv is a non-profit, community-oriented organization. Solar is a multi-trillion dollar finance opportunity. Is crowd funding for real? Can a participatory, community-oriented movement disrupt energy?

# Participatory Finance: Wind in Denmark

The energy industry is a hierarchical, command-and-control world. Big Banks invest in Big Energy Assets that Big Utilities operate to sell energy to individuals, families, and businesses. Energy flows one way (from Big Energy to the user) and cash flows the opposite way (from the user to Big Energy). All the decisions in Big Energy are made by a handful of individuals and committees that for the most part are not accountable to users or to society at large.

Participatory energy is when individuals and families participate in the generation, transmission, and storage of their own energy and their community's energy.

Participatory finance is when individuals and families invest directly in the energy assets that they or their extended communities will use. In participatory finance, individuals invest directly in energy assets. They choose which energy projects (clean, distributed, relatively small) they want to invest in and they profit from the cash flows that these projects generate.

Participatory energy and participatory finance can go hand in hand, as Denmark showed in the 1970s. Denmark's wind energy market commands the largest national market share in the world.

On November 3, 2013, wind generated over 100 percent of the power demand in Denmark, breaking a world record.[85] But Denmark's was not your ordinary command-and-control energy market. The country with the largest wind installed base in the world got that way essentially without investments by or the participation of the country's largest power utilities.

Denmark got to be the world's largest wind market because communities invested in their own wind-generation assets.

To encourage communities to adopt wind energy, the Danish government created incentives for individuals and families to invest in wind power in their own communities. As the market grew and turbines became larger and required larger capital investments, individuals would invest in shares of "wind turbine cooperatives." These cooperatives would in turn develop and invest in community wind turbines and wind farms.

By 2001, over a hundred thousand families belonged to wind turbine cooperatives. These cooperatives had installed 86 percent of all wind turbines in Denmark.[86] Wind cooperatives in Denmark present the first important example of participatory finance in energy. They also present the first important example of participatory energy. In Denmark, individuals chose what type of energy they wanted to use (usually wind, but also geothermal or solar) and share with the community. Denmark is also the first important example of what it means to transition from large, centralized power plants to the distributed generation model.

Participatory finance and participatory energy went hand in hand in Denmark. By 2005, over 150,000 families owned turbines or were members of turbine cooperatives, and these cooperatives provided 75 percent of all wind power in Denmark. At that point the private sector started catching on to the importance of distributed wind power.[87]

By 2008, Denmark's wind power capacity had increased such that wind now provides 19.1 percent of the country's electricity (see Figure 2.2).[88]

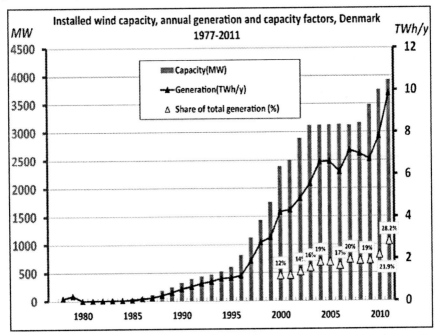

*Figure 2.2—Installed wind capacity in Denmark, annual generation and capacity factors, 1977-2011. (Source: Wikipedia)[89]*

With 86 percent of Danes supporting clean distributed energy, Denmark's wind market kept growing. At the end of 2012, fully 30 percent of the country's electricity was supplied by wind (see Figure 2.2). The country plans to generate 50 percent of its total demand from wind by 2020.

Denmark's participatory energy model spread to the rest of Europe. Germany caught on in a big way. Germany has the world's largest installed base of solar power and it got that way without the participation of its large power utilities.

Germany's solar power was a case of participatory energy; it was mostly individuals and small businesses that chose to develop and install solar power plants on their premises. These solar installations were mainly financed by the banks, so Germany did not have the participatory finance element that was so important in Denmark.

Smaller crowd-funding (participatory finance) companies have sprung up in places like the United Kingdom and Holland to raise money for wind and solar projects.

On September 24, 2013, WindCentrale announced that it had raised €1.3 million to build and operate a wind power plant — but this was not your typical fundraising campaign.[90] The company sold 6,648 shares at €200 each to individuals in 1,700 households who, by buying shares, would own shares in a 2MW wind turbine. Each share would get the profits from the generation of 500 kWh per year. According to WindCentrale, shares would yield 8.5 percent annually, assuming that energy costs go up by 3 percent annually.[91] The energy will be sold through a company called GreenChoice that sells clean electricity to more than 350,000 customers in Holland.

Can the United States catch up with its European cousins? Who will finance the multi-trillion dollar solar opportunity here? Will it be big banks (the case in Germany) or individuals and families (the case in Denmark)?

## The Golden Gate Bridge as an Example of Participatory Finance

Participatory finance and crowd funding are seen as a new phenomenon, but not to San Franciscans. They have been doing it for decades. The Golden Gate Bridge, perhaps the most iconic and loved human-made structure on the west coast of the United States, would not have been built without the participation of the people of California.

When the Golden Gate Bridge was proposed in 1916, San Francisco was the largest American city still being served by ferryboats. Joseph Strauss, an ambitious engineer and entrepreneur, spent the better part of a decade drumming up support for the bridge in Northern California. At the time, the Golden Gate Ferry Company was the largest ferry operation in the world. It was also a subsidiary of one of the most powerful (financially and politically) businesses of the time: the Southern Pacific Railroad.[92]

The ferry business was a cash cow that would be disrupted if a bridge were built across the Golden Gate.

In 1928, the California legislature passed the Golden Gate Bridge and Highway District Act. It created a special-purpose district of six counties (San Francisco, Marin, Sonoma, Del Norte, and portions of Napa and Mendocino) with the goal of financing the design, construction, and operation of the bridge.[93] The bridge was expected to cost $35 million. After the Wall Street crash of 1929 plunged the nation into the Great Depression, public funding for the bridge dried up.

Strauss and his supporters decided to raise the money directly from the people of California. On November 4, 1930, voters within the Golden Gate Bridge and Highway District's six member counties went to the polls to decide the question of whether to put up their homes, farms, and business properties as collateral for a $35 million bond issue to finance the construction of the Golden Gate Bridge.

Southern Pacific, the ferry lobby, and their allies orchestrated a vicious misinformation campaign. It was a classic fear uncertainty and doubt (FUD) campaign similar to campaigns that the fossil fuel and nuclear industries would use later against clean energy. Opponents of the bridge argued:

> The bridge clearance would prevent the world's great ships from entering the Bay. An enemy fleet could demolish the bridge and bottle-up the U.S. fleet. The bridge could not be built. It would not stand. The floor of the Golden Gate Strait would not support the weight of the San Francisco pier and tower. The entire project was a hoax and sham. Only fools would buy bonds of a bridge certain to fall. Taxpayers would suffer and have to continue paying to finance the fiasco.[94]

After the votes were counted, it was clear that the people believed in Chief Engineer Joseph Strauss's vision. They voted 145,057 in favor and 46,954 against building the Golden Gate Bridge.

These were no ordinary votes. The people of Northern California voted (by a 3 to 1 margin) to put up their own properties as collateral to build the Golden Gate Bridge. With the collateral of the people, Bank of America pledged to purchase the first $3 million block of bonds to commence bridge construction in 1932.

It was the financial and political participation of the people of the San Francisco Bay Area that made the Golden Gate Bridge possible. The bridge opened in May 1937. In its first full year of operation, 3.3 million vehicles crossed the bridge. By 1967, the number of vehicles crossing the Golden Gate Bridge grew to 28.3 million.[95]

Ferry service between San Francisco and Marin County dwindled until it officially ended in 1941. The ferry industry was officially disrupted.

## Mosaic: A Company Devoted to Participatory Finance

I was thinking about the Golden Gate Bridge story as I stood on the deck of the ferry taking me from the Ferry Building in San Francisco to Jack London Square in Oakland. After living in San Francisco for twenty years, I still love to experience the San Francisco Bay from the deck of the ferry. Ferries made a comeback in the 1970s and now fill a small niche market supplementing the Bay Area transit system.

I am an advisor to the SFUNCube solar accelerator, located near Jack London Square in Oakland. This area reminds me of Kendall Square in Cambridge, Massachusetts, next to my alma mater MIT, before it became a global hub for technology startups. The Ferry landing in Oakland is near a beautiful marina, shops, and restaurants, but walk a few blocks east and you come to shuttered warehouses that show the scars of the latest wave of industrialization.

As an advisor to SFUN, I review, with co-founders Emily Kirsch and Danny Kennedy, business plans that get submitted by applicants around the world. SFUN believes that software and financial innovations are keys to the next wave of solar entrepreneurship. Few companies hold both these keys. Mosaic is a solar crowd-funding company located in SFUNCube's facilities.

Dan Rosen, Mosaic's co-founder and president, told me that solar is a trillion-dollar opportunity that will be served best by a peer-to-peer market. When people invest directly in a solar project, they benefit through the interest rate they earn; the users of the solar power plant, meanwhile, benefit from lower energy bills.

Mosaic launched in January 2013. By the time Dan Rosen and I spoke in December of that year, three thousand people had invested more than $6 million in 25 solar projects. The projects ranged from a 55 kW solar installation at a low-income housing complex in Corte Madera, California, to a 1.6 MW system in Prairie View Solar Park in Gainesville, Florida.[96]

Mosaic sees its mission as one of democratizing access to the multi-trillion dollar solar investment opportunity.

Fewer than twenty banks have consistently participated in solar in the U.S. This concentration of capital means three things:

1. Capital for solar projects is limited. Even if the twenty banks (or so) had a healthy appetite for solar, portfolio theory dictates that they limit their exposure to any single asset class.

2. The returns banks demand is high and not commensurate to the adjusted risk/reward for solar. The Federal Deposit Insurance Corporation (FDIC) insures deposits at nearly 6,800 institutions[97]. Fewer than twenty of them invest in solar. Despite the high liquidity and low interest rates in financial markets, the fact that only a few banks participate in solar finance means that there is a lack of competition among banks. This pushes up supplier power and the cost of capital for solar.

3. Transaction costs are high. There are no industry-accepted project investment standards. Every project seems to use a different set of power purchase agreements and other documents. This pushes up the fees that lawyers charge and the time it takes to process and close investments. "Do we really have to pay lawyers $70,000 for every solar project?" Dan Rosen asked.

Most of the projects listed on the Mosaic website provide yields to investors in the 4.5- to 5.75-percent range; two outliers yield 7 percent. For individual investors who are getting less than 1 percent on their bank accounts or certificates of deposit (CDs), a return over 4 percent is considerable. So far, Mosaic projects have 100-percent on-time payments. There is no reason for the credit card-type returns that banks have so far demanded of solar project developers.

Financial services companies and energy companies have huge appetites for outsize returns on capital. A peer-to-peer technology platform like Mosaic removes financial services companies and energy companies from the playing field. When small investors participate in a peer-to-peer platform like Mosaic, both sides win; individual investors benefit from a stable, long-term cash flow with a return that was previously available only to the large players in energy, while solar users benefit from the cheaper, more stable energy flow. "For every 100 basis points (1 percent) we reduce in the total cost of capital, the cost of solar electricity drops by 1 ¢/kW to 2 ¢/kWh," reported Dan Rosen, president of Mosaic. "As solar approaches and beats grid parity around the nation, a 1-percent savings in the cost of capital makes a huge difference."

Dan Rosen sees Mosaic as a means of expanding opportunity. Mosaic gives solar end users and smaller investors access to opportunities they would otherwise not have in today's energy finance paradigm.

"It's astounding how inefficient the solar finance and development process is," said Rosen. "We're trying to bring standardization to solar finance. Our goal is to make solar loans be like auto loans: you apply online and get instant approval for your solar project."

Mosaic wants to build an Internet cloud-based platform where developers of solar projects of any size can post their projects and investors of any size can invest in them. The projects and investors can be located anywhere. Said Rosen, "The small investor could put a few hundred dollars and the pension fund can put a few million. Community-oriented investors can find and invest in projects in their own neighborhoods while investment managers can build a portfolio of several projects in different markets."

About a hundred years ago, GMAC created auto loans; it has made hundreds of billions of dollars since. Along the way GMAC helped turn the early-stage automotive industry into a multi-trillion dollar powerhouse. Mosaic's ambitions are to do with solar what GMAC did with autos by using participatory finance.

"Americans have five trillion dollars in their IRAs," said Rosen, "What we want to do is give every American the opportunity to say, 'I have a solar farm in my IRA account.'"

Wall Street eventually caught up to GMAC. However, technology platforms have network effects, and this makes a winner hard to catch up with. Maybe Mosaic is the first trillion dollar financing opportunity enabled mainly by a peer-to-peer platform. Will Wall Street miss out on this multi-trillion dollar energy finance opportunity?

## Why Warren Buffett and Wall Street Like Solar

In February 2012, two months after Warren Buffett's MidAmerican Energy Holdings acquired the largest solar power plant in the world for $2.4 billion, it went to Wall Street to refinance the project. MidAmerican needed Wall Street's help to raise the first $850 million tranche. Despite the backing of Warren Buffett, the credit rating outfits were not crazy about solar. Fitch Ratings gave the bond offering a BBB- rating, its lowest investment-worthy credit rating. Moody's rated it Baa3; Standard & Poor's assigned it a BBB- rating, its lowest investment grade rating, according to Bloomberg.[98]

Topaz Solar Farms LLC offered the $850 million unsecured debt with a maturation date of Sept 2039 and a yield of 5.75 percent per year. Equivalent U.S. treasuries yielded only 1.95 percent, according to Bloomberg. Buffett's solar deal offered almost three times the return of U.S. government treasuries. Investors snapped it up. The Topaz Solar bond offering was oversubscribed by $400 million.

Investors who want a steady, stable, long-term return have nowhere to go these days. In August, the 1-year US Treasury bond returned only 0.13 percent;

the 5-year yield returned only 1.47 percent.[99] Stocks potentially have a better upside but are extremely volatile. The U.S. has seen two traumatic stock market crashes since 2000. The Nasdaq Composite (IXIC), for example, went from an adjusted high of 4,571 in January 2000 to 1,172 in July 2002; it rose to 2,700 in July 2007 only to crash again to 1,528 in January 2009. At the end of 2013, the Nasdaq was back up to 4,177 (see Figure 2.3).

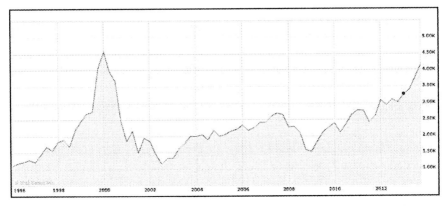

*Figure 2.3—The Nasdaq Composite. (Source Yahoo! Finance, © Yahoo! Inc.)[100]*

Mortgage securities? Many still remember the mortgage collateralized debt obligations (CDO) that helped trigger the financial crisis (a crisis the U.S. is still recovering from). Commodities are more like roller coaster rides.

The summer of 2013 may be remembered as the year public markets discovered solar and wind. The summer of hot sunny love, if you will.

Recall that Warren Buffett's company, MidAmerican Energy Holdings, invested $2.4 billion to acquire the Topaz solar development project. Many books have been written about Buffett's investment wisdom. Why did he invest in Topaz? For the answer, consider some of Buffett's fundamental rules of value investing. According to Warren Buffett's "Market Wisdom":
  • "I like businesses I can understand."
  • "I want to know what a business will look like ten years from now. If I can't see them where they will be ten years from now, I don't buy them."
  • "We don't have huge returns, but we don't lose our money either."[101]

In other words, Warren Buffett likes to buy easy-to-understand, boring businesses that can generate good cash for decades and, in the worst case scenario, don't lose the money he put into them.

MidAmerican Energy bought a solar power plant, not a solar photovoltaic technology provider. There's a great deal of difference between these two businesses and the media doesn't usually distinguish between the two.

The economics of a solar power plant project are boring. As in any business, the net income from a solar power plant is equal to revenues minus costs. Assuming you have a power purchase agreement (PPA), this is what revenues and costs look like:

- Revenues: power demand * power prices
  - o Power price—Stable (fixed) over the life of the power purchase agreement (20 years or so).
  - o Power demand—Stable over the life of the PPA.
- Costs: fuel costs + O&M costs + insurance costs + cost of capital
  - o Fuel costs—Zero. Sunshine is free and the sun is expected to shine for another billion years or so.
  - o Operations & maintenance (O&M) costs—Stable and extremely low (less than 1 ¢/kWh).
  - o Insurance costs—Usually a low percentage (0.3 percent) of the cost of assets, which decrease over the life of the power plant.
  - o Cost of capital—Basically, the cost of interest, which is dependent on the interest rate.

Like a home mortgage, the cost of capital has become the most expensive line item in the construction of a solar power plant. The lower you can push down the interest rate, the lower the cost of solar electricity will be.

As the installed cost of solar keeps decreasing, the cost of solar electricity will approximate the cost of capital. This is good news because the cost of capital for the U.S. government is less than 1 percent. The cost of capital in Japan has been essentially zero for more than a decade.

It's been a long dating period, but Wall Street is starting to understand the value of solar and the wedding will soon take place. Large solar power plants are benefiting from the lower costs of capital. This will allow them to lower the cost of the solar electricity that they deliver.

The solar virtuous cycle is in motion. The lower cost of solar leads to increased market adoption, and this, in turn, lowers the perceived risk and attracts more capital at a lower cost for capital. And this, in turn, lowers the cost of solar, which leads to increased market adoption, not to mention increased investment, more innovation, and even lower costs for capital. Once this virtuous cycle reaches critical mass, market growth will accelerate. Solar will become unstoppable and the incumbents will be disrupted.

But what about financing smaller distributed residential and commercial solar installations? Wall Street deals take time and the transaction costs are high. Wall Street investment bankers prefer large deals rather than a bunch of smaller residential deals. How will distributed solar tap the trillions it needs from Wall Street?

## Securitization Comes to Solar

On November 13, 2013, SolarCity announced an offering of $54.4 million in solar asset-backed notes.[102] The notes would have an interest rate of 4.8 percent and a maturity date of December 21, 2026. Standard and Poor's gave the offering (titled SolarCity Series I LLC Series 2013-1) a BBB+ rating.[103] As shown by a New York Times table of select corporate bond offerings (see Figure 2.4), S&P put SolarCity's 2013-1 bond credit rating a bit higher than Ford Motor's F.GY (rated BBB-) and below AT&T's T4013485 bonds (rated A-).

**FINRA TRACE Corporate Bond Data**

INVESTMENT GRADE   HIGH YIELD   CONVERTIBLES                                    12/27/2013

| | | | Credit rating | | | Price | | |
|---|---|---|---|---|---|---|---|---|
| Issuer name (symbol) | Coupon % | Maturity | Moody's | S&P | Fitch | Last | Change | Yield % |
| General Elec Cap Corp Medium Term Nts Bo GE.ATZ | 2.15% | Jan '12 2015 | A1 | AA+ | n.a. | 101.271 | −0.61 | 0.90% |
| Continental Res Inc CLR3875305 | 5.00% | Sep '12 2022 | Baa3 | BBB- | n.a. | 103.5 | −0.125 | 4.35% |
| Ford Mtr Co Del F.GY | 7.45% | Jul '12 2031 | Baa3 | BBB- | BBB- | 121.75 | −1.23 | n.a. |
| Procter & Gamble Co PG.HT | 1.80% | Nov '12 2015 | Aa3 | AA- | n.a. | 102.551 | +0.183 | 0.43% |
| Devon Energy Corp New DVN3852854 | 3.25% | May '12 2022 | Baa1 | BBB+ | BBB | 95.711 | −0.255 | 3.85% |
| Anheuser Busch Inbev Worldwide Inc BUD.IZ | 7.75% | Jan '12 2019 | A3 | A | A | 125.518 | +0.424 | 2.35% |
| General Elec Cap Corp Medium Term Nts Bo GE3672800 | 1.63% | Jul '12 2015 | A1 | AA+ | n.a. | 101.611 | +0.109 | 0.55% |
| At&t Inc T4013485 | 4.30% | Dec '12 2042 | A3 | A- | A | 85.608 | −2.96 | 5.26% |
| Apple Inc AAPL4001809 | 2.40% | May '12 2023 | Aa1 | AA+ | n.a. | 91.607 | −0.08 | 3.46% |
| Commonwealth Bk Australia Medium Term Nt CBAU3828562 | 2.25% | Mar '12 2017 | Aaa | n.a. | AAA | 102.934 | −0.225 | 1.31% |

*Figure 2.4—FINRA TRACE Corporate Bond Data, Dec 27, 2013, New York Times. (Source: NYTimes.com)*

This bond offering, small as it was, represented a breakthrough for the U.S. solar industry. Securitization is a major step forward in increasing the capital flow to an industry. Securitization also lowers the cost of capital to an industry.

SolarCity's securitization deal pooled 5,033 residential solar photovoltaic system contracts into a single security. When the 5,033 residential customers pay their monthly electricity or lease rates, their payments will be aggregated and paid to investors in the security. Who invests in solar securities? Pension funds, university endowment funds, and other investors who favor predictable long-term cash flows. Solar City benefits by redeploying $53 million of fresh cash to build thousands of new residential solar systems.

The SolarCity deal is a small first step in opening up the $1.8 trillion asset-backed securities industry to solar.[104] In time, pension funds and investment banks will get comfortable with solar-backed securities. They will ask leading solar installers such as SolarCity, SunRun, and Sungevity to sell them more solar asset-backed securities. As more buyers (pension funds, endowment funds, investment banks) and more sellers (third-party solar installers) participate in this process, the market will become more liquid; more cash will flow into the market. Consequently, the cost of capital for high-quality solar asset-backed securities will go down.

The availability of fresh capital allows the solar market to scale up more quickly. Meanwhile, the lower cost of capital lowers the cost of solar electricity. More customers can buy solar because it's cheaper; companies can build solar systems at a quicker pace. Installers can then securitize thousands of residential solar loans and raise more money at lower capital costs.

A century ago, widely available credit and the lower cost of capital created a virtuous cycle in the auto industry. As more new consumers had access to credit, more cars were sold, and as more cars were sold, the auto industry experienced more investment, innovation, and economies of scale. This, in turn, lowered the sticker price of cars, which brought still more consumers into the market, which brought new banks into the credit market, which lowered capital costs again, which brought even more consumers into the market, and so on. The end result was a virtuous cycle in the auto industry.

Just as securitization helped democratize access to autos, homes, and a college education, solar securitization has the potential to democratize access to solar energy. The solar virtuous cycle is thus further accelerated by the creation of solar asset-backed securitization.

## Financing Solar with Real Estate Investment Trusts (REITs)

In April 2013, Hannon Armstrong Sustainable Infrastructure Capital, Inc. went public on the New York Stock Exchange (NYSE: HASI). It sold 13.33 million shares and raised $155.4 million.[105] The IPO was otherwise unremarkable except for one important fact: HASI was a real estate investment trust (REIT) focusing on clean energy investments.

A real estate investment trust (REIT) is a legal structure by which a company can invest in, own, and operate income-producing real estate.[106] REITs were created by the U.S. Congress in 1960 to give investors the opportunity to invest in real estate in the same way they can invest in liquid market securities such as stocks and bonds.[107] Since 1960, REITs have dramatically expanded in scope and geography. Being a REIT gives Hannon Armstrong Sustainable Infrastructure Capital enormous advantages when it comes to financing clean energy projects. "REITs have a market valuation of $630 billion and give average dividends of 5 percent," according to Dan Reicher, a professor at Stanford University who is executive director of Stanford's Steyer-Taylor Center for Energy Policy and Finance. "Clean-energy REITs would have access to hundreds of billions of private investor dollars at far lower capital costs than they are currently getting."

Who decides whether an asset class such as solar or wind qualifies as a REIT? The Internal Revenue Service (IRS) does. By means of a "revenue ruling," the IRS can decide what types of asset a REIT can invest in. REITs can now invest in office buildings, apartment buildings, warehouses, shopping centers, and hospitals. Even timber investments are "REIT-able" (the term for asset types that qualify for REIT investments).

Earlier in this chapter, I explained that one requirement of PACE financing is that the asset be attached to the property. Rooftop solar PV plants are attached to the property (they are attached to apartment buildings, warehouses, and homes). Solar qualifies for PACE financing but the IRS has not qualified solar as a REIT-able investment.

Can investors convince the IRS that solar, because it is attached to property, can qualify for REIT investments? A company can request, and the IRS can issue, a "private letter ruling" that details whether a type of infrastructure qualifies for REIT investments. Hannon Armstrong got such a "private letter ruling" from the IRS.

In his testimony to Congress in October 2013, Dr. Reicher of Stanford mentioned how a company like FirstWind told him that they pay up to 14-percent cost of capital for raising tax equity.[108] To put that into context, Hannon Armstrong's dividend yield was 3.19 percent, according to Morningstar.[109] Because Hannon distributed 100 percent of its earnings, its dividend yield is effectively its cost of capital.

Being able to raise money directly from investors by using an instrument such as a REIT could lower clean energy costs substantially. Imagine the difference between monthly payments made on a home mortgage with a 14-percent interest rate and a home mortgage with a 3.19-percent interest rate. That is the difference between using a REIT and using conventional financing.

On December 23, 2013, Hannon Armstrong announced that it had sold $100 million of asset-backed sustainable yield bonds™ (HASI SYBs) with an even lower, 2.79-percent yield.

According to Hannon Armstrong CEO Jeff Eckel, this $100 million transaction "securitizes the cash flows generated by over 100 individual wind, solar and energy efficiency installations."[110]

Making clean energy assets REIT-able could cut the cost of solar (and wind) electricity as much as a third, according to a letter that 35 members of Congress sent to President Obama in December 2012. The letter read in part:

> Small tweaks to the tax code could attract billions of dollars in private sector investment to renewable energy deployment, reduce the cost of renewable electricity by up to one third, and dramatically broaden the base of eligible investors.[111]

It actually doesn't take an act of Congress to make solar and wind REIT-able, according to Dr. Reicher of Stanford. All it takes is an administrative "revenue ruling" by the IRS. Hannon Armstrong may have opened the door for the IRS to move from a "private letter ruling" to a "revenue ruling."

The congressional letter to President Obama also urged him to open up another type of legal structure to clean energy: master limited partnerships.

## Extending Master Limited Partnerships
## to Clean Energy

Kinder Morgan (NYSE: KMI) was founded in Houston in 1997 when Richard Kinder, William V. Morgan, and a group of investors acquired Enron Liquids Pipeline, LP, a small, publicly traded pipeline company.[112] Kinder Morgan now owns or operates 82,000 miles of oil and gas pipelines and 180 terminals that store products such as oil, coal, and petroleum coke.[113] By using a business structure known as a master limited partnership (MLP), Kinder Morgan has built an energy giant valued at $102 billion. KMI is the fourth largest publicly traded energy company in North America based on enterprise value.[114]

Kinder Morgan is not the only energy company to use MLPs. Apache Petroleum used a master limited partnership (MLP) structure to go public as early as 1981. Since then, energy MLPs have raised more than $400 billion for building oil and gas pipelines, drilling, mining, transporting energy, and processing oil, gas, and coal.

As a business structure, the master limited partnership has the tax advantages of a limited partnership, but its stock can be traded like corporate stock. Unlike typical "C" corporations, which have to pay corporate taxes, MLPs pay no corporate tax. This gives them a big advantage.

Their net earnings pass through to the shareholders as dividends. (See Figure 2-5.)

Kinder Morgan and the recent "shale" oil and gas boom in the United States could not have happened without MLPs. MLPs allowed the oil and gas industry to raise money from public investors in spite of the financial downturn of the late 2000s.

Master limited partnerships cannot be used for clean energy development projects. Congress does not allow MLPs to be used for "inexhaustible" natural resources such as wind and solar. This is yet another way in which the U.S. government gives the fossil fuel industry a distinct institutional advantage over clean energy.

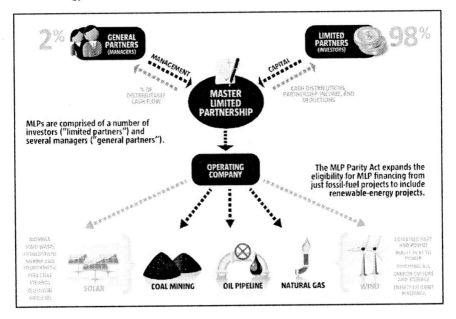

*Figure 2.5—How master limited partnerships (MLPs) work. (Source: U.S. Senator Chris Coons)[115]*

Dan Reicher, Stanford University professor and former Assistant Secretary of Energy, has proposed using MLPs for clean energy projects. He believes clean energy MLPs could substantially increase the number of investors and decrease the cost of capital for clean energy projects.[116]

According to Reicher, MLPs have a market capitalization of $440 billion and pay on average a dividend of 6 percent. This compares to the 10- to 20-percent "credit card-like" cost of capital that clean energy companies have to pay for tax equity.

To understand the difference that MLPs can make, start by considering why it's hard to raise equity capital. The main incentive program for solar in the U.S. is the investment tax credit (ITC).

The ITC is a 30-percent tax credit for solar systems on residential and commercial properties.[117]

Oil and gas companies also receive investment tax credits. For example, they receive the foreign tax credit deduction and the deduction for intangible drilling costs. Respectively, these two tax deductions will save the five largest oil companies $2 and $7.5 billion over the next decade, according to the Congressional Joint Committee on Taxation.[118]

Individual projects cannot directly use the ITC until they start producing profits, which can take several years. By contrast, large fossil fuel energy companies already have huge profits and they can offset these profits right away by using investment tax credits.

Solar companies in the United States raise equity capital by attracting investors who can use investment tax credits. This is called "tax equity." There are several issues with using tax equity to raise capital for a project. The first issue is that this is a highly illiquid market. In any given year only ten to twenty investors nation-wide have the appetite for the billions of dollars of tax equity needed by solar developers. The second issue is that the lack of competition allows these few equity investors to charge "credit card" rates for their capital.

If the oil and gas industry only had access to the limited number of investors that the government allows solar and wind to access, the industry could never have developed the millions of wells and thousands of miles of pipelines it did over the last decade.

Master limited partnerships would give solar and wind companies the opportunity to directly access millions of investors through public capital markets. Liquidity alone would cut the cost of capital for clean energy projects.

The Master Limited Partnership Parity Act, a bill introduced in the U.S. Congress in 2012, would extend the use of MLPs to clean energy. It would level the playing field. It would allow other forms of energy generation and energy-efficiency projects to take advantage of master limited partnerships. Wind, solar, gasification, waste-to-energy, carbon capture, as well as energy efficiency building projects would all be able to take advantage of MLPs.

The Master Limited Partnership Parity Act was introduced in Congress by Senators Chris Coons (Democrat of Maryland) and Jerry Moran (Republican of Kansas). In 2013 it was amended to broaden the scope of energy projects that it covers. The bill has been "referred to Committee on Finance" and is still waiting for a hearing in Congress.[119]

The Joint Committee on Taxation studied the financial impact of extending the Master Limited Partnership Parity Act to clean energy. It concluded, "The MLP Parity Act is a bargain." The Committee said the Act could result in $10 billion in clean energy investments "right away."[120] Moreover, the Act would cost taxpayers $307 million over five years and $1.3 billion over ten years compared with a forecasted taxpayer cost of $6.7 billion over ten years for existing fossil-fuel MLPs.

The Master Limited Partnership Parity Act "is less filling and tastes great," to quote the old beer commercial. It would lower the cost of energy to consumers, create new jobs, and lower the tax bills. What's not to like?

## In Conclusion: The Trillion Dollar Solar Finance Opportunity

Solar finance has come a long way since 2008 when the "no money down," solar-as-a-service business model was introduced. Owing to experimentation with new forms of finance and innovations in existing securities, a niche, illiquid tax-equity investing market has quickly increased its capital base and decreased its capital costs. (See Figure 2.6.)

Americorp Energy Holdings, owned by Warren Buffett's Berkshire Hathaway, is now the largest solar developer in America. The company has acquired two of the largest solar power development projects in the world. Topaz Solar Farms LLC's offering of $850 million unsecured debt with a 5.75-percent yield successfully closed. In fact, it was oversubscribed by $400 million.

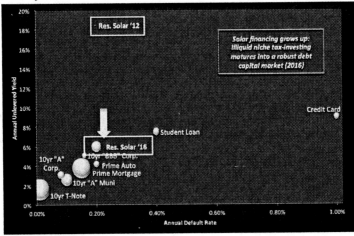

*Figure 2.6—Residential solar finance cost of capital evolution from 2012 to 2016. (Source: Clean Power Finance)*

Mosaic, based in Oakland, California, a crowd funding platform, became the first solar finance company in the U.S. to successfully engage in for-profit participatory finance for solar projects. These projects are being financed for about 5 percent cost of capital. Mosaic is also building an Internet platform that it hopes will democratize energy finance. Individuals will be able to invest directly in, and profit from, the multi-trillion dollar solar infrastructure.

In November 2013, SolarCity offered the first residential solar securitization deal in history. The deal opened the door to a more liquid residential solar finance market. The SolarCity deal was for $54.4 million in solar asset-backed notes at a 4.8-percent interest rate.[121]

The following month, Hannon Armstrong, the first investment fund to successfully go public as clean energy REIT, announced that it sold $100 million of asset-backed sustainable yield bonds™ (HASI SYBs) with an even lower yield: 2.79 percent.

Kristian Hanelt of San Francisco-based Clean Power Finance (CPF) expects the solar finance market to become a robust capital market by 2016. Clean Power Finance is a financial services and software company that manages a half-billion dollars of residential solar project financing capital.[122] The CPF business model takes advantage of Internet cloud-based software tools. Using these tools, investors and lenders can access and invest in residential solar projects managed by solar installers who also use the CPF website.

Since 2006, residential and commercial solar in the U.S. have grown at a 76-percent annual rate.[123] Like the financial and business model innovations that GMAC pioneered in the early 20th Century — innovations that propelled the auto industry to become America's most important manufacturing industry — solar finance and business model innovations are tapping into private capital as never before. This capital will finance the trillion-dollar solar disruption of the world's largest industry.

# Chapter 3:

## Electricity 2.0.

## Distributed, Participatory Energy and the Disruption of Power Utilities

*"You never change things by fighting the existing reality.*
*To change something, build a new model that makes the existing*
*model obsolete."*

*- Buckminster Fuller*


*"Cellular phones will absolutely not replace local wire systems."*

*- Martin Cooper, co-inventor of the first handheld mobile phone (1981)*


*"Change before you have to."*

- Jack Welch, former CEO, General Electric

On November 5, 2012, the city of Palo Alto, California, announced a 25-year deal to buy solar power for about 7.7 ¢/kWh.[124] By comparison, PG&E, the largest utility in northern California, charges a minimum of 13 ¢/kWh (baseline) and a maximum of 34 ¢/kWh for "Tier 4" and "Tier 5" usage (see Figure 3.1). For solar power, Palo Alto pays about half the minimum price that PG&E charges its residential rate payers; the city pays about a fifth of what PG&E charges for higher usage consumption (read: summer air conditioning).

The shock of disbelief hadn't yet settled when, seven months later, Palo Alto announced another agreement to purchase solar power for even less: 6.9¢/kWh.[125]

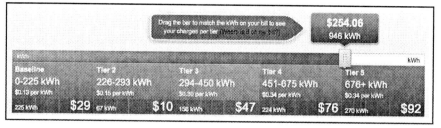

Figure 3.1—PG&E rates for San Francisco, CA. (Source: PG&E website)[126]

Palo Alto has a 100-percent clean energy mandate and is on track to buy 30-percent clean energy by 2015 and 48-percent clean energy by 2017. These numbers don't include solar that homeowners and businesses have installed on their rooftops. Not only will Palo Alto be powered with inexpensive and 100-percent clean electricity, the low costs it pays for electricity are guaranteed for the next 20 to 25 years.

As solar technology improves, the market scales and financing costs decrease. Solar generation costs are decreasing quickly. The distributed nature of solar makes the disruption of the existing utility business model inevitable. This disruption will happen faster than existing command-and-control energy companies expect.

Existing energy companies are missing the big picture. Every single aspect of solar is distributed: technology innovation, design and development, finance, installation, and maintenance. Some pundits expect many decades to pass before the market adopts solar energy, but solar markets, because of their distributed nature, can turn on a dime.

Most utilities have responded to the distributed clean energy disruption by hiring lobbyists, lawyers, and accountants to justify charging higher rates and new fees. What these "captains" of the energy industry are doing is tantamount to raising the price of food on the Titanic.

Higher rates and new fees may increase their short-term cash flow but they won't prevent the inevitable disruption.

## Australia: The Shape of Things to Come

In 2008 Australia had barely any solar at all. By the end of 2012 it had already crossed the threshold of one million solar homes (see Figure 3.2).[127] Australia went from essentially nothing to a penetration of more than 11 percent of the residential power market in about four years.

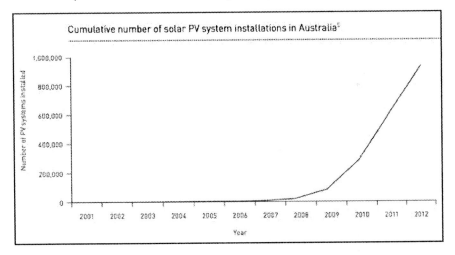

*Figure 3.2—Number of solar PV system installations in Australia. (Source: Australia Clean Energy Council)*[128]

To put these numbers in perspective, compare Australia with the state of California and the United States. Australia's population is 23 million people.[129] California has 38 million people and the U.S. has 313 million.[130]

California is the largest solar market in the United States. While market share fluctuates from year to year, the Golden State has historically represented about half the U.S. solar market. According to the California Solar Initiative, there were 167,878 solar installations at the end of 2012. This number includes residential, commercial, and large-scale power plants.[131]

If the state of California had Australia's 11-percent solar penetration, there would be 1.65 million solar installations, nearly ten times more than there are today. If the United States had a residential solar penetration equivalent to Australia's, there would be 13.6 million solar homes. The actual number of solar installations in the U.S. at the end of 2012 was 300,000, or about 2 percent of where it would be if the U.S. matched Australia's rate of solar penetration.[132]

Furthermore, the 11-percent penetration rate is an average number for Australia. In the state of South Australia, a full 20 percent of homes have solar rooftops. Some neighborhoods have a solar penetration of 90 percent, according to Mike Swanton of Energex, a Queensland utility.[133]

What happens to a power utility when users start generating their own solar energy?

1. Demand for utility energy drops. As users generate their own energy, they buy less from the utility.
2. Competition increases. The utility competes with myriad solar installers.
3. Utility revenues drop. As demand drops and competition increases, the utilities make less money.
4. Utility margins drop disproportionately. Solar generates the most energy during peak pricing billing cycles, which undercuts the power utility's highest margin products.

Australian electricity prices have gone up 50 percent over the last five years, from about 25 ¢/kWh to 38 ¢/kWh.[134] The price increases occurred despite the fact that Australia is a major coal and natural gas producer. In 2103, solar was already as low as 12 ¢/kWh — and dropping.

Said Jack Welch, former CEO of General Electric, "If the rate of change on the outside is greater than the rate of change on the inside, the end is near." The utility business model is obsolete and, for utilities, the end is near.

## How Solar Disrupts Retail Peak Pricing

Most of the conversation in solar is about reaching the magic "grid parity," also called "socket parity." Grid parity is what happens when an alternative energy source generates electricity at the same price as power from the electric grid. However, achieving grid parity is only part of the reason solar is disrupting traditional utilities.

At the retail level, distributed solar generation is disruptive to the conventional utility business model because it destroys its most lucrative revenue stream: peak pricing.

Utilities have historically made a disproportionate amount of income from something called peak prices. Arizona Public Service, for example, may charge just 5 ¢/kWh during off-peak hours, but nearly five times as much (24.4 ¢/kWh) during on-peak hours, and nearly ten times as much (49.4 ¢/kWh) during the "super-peak" hours of June, July, and August, when Arizona is at its hottest (see Figure 3.3).[135]

It turns out that Arizona Public Service (APS) charges almost ten times its "base" rate when the sun is shining brightest. A residential user who generates her own solar energy starts saving money on day one because the cost of rooftop solar is much lower than the  peak prices that her utility charges.  In Arizona, with its year-round sunshine and sunny summers, distributed rooftop solar PV is already much cheaper than what APS charges its ratepayers.

As more customers adopt solar and purchase less electricity from the utility at peak prices, the utilities' high premiums start to disappear. Utilities can't turn their large nuclear or coal "baseload" power plants on and off with market demand. These outdated plants keep producing power whether the demand is great or small.

Utility executives have taken notice of how quickly and dramatically peak premiums have dropped in markets with high penetration of distributed solar. The peak premium in Germany dropped by 80 percent in just five years, from €14 per MWh in 2008 to just €3 per MWh in 2013, according to the Fraunhofer Institute for Solar Energy Systems.[136]

| Energy Charge: | | |
|---|---|---|
| | June – August Billing Cycles (Super Peak Summer) | |
| | $0.49445 per kWh during Super-Peak hours, plus $0.24445 per kWh during On-Peak hours, plus $0.05254 per kWh during Off-Peak hours | |
| May, September, and October Billing Cycles (Summer) | | November – April Billing Cycles (Winter) |
| $0.24445 per kWh during On-Peak hours, plus $0.05254 per kWh during Off-Peak hours | | $0.19825 per kWh during On-Peak hours, plus $0.05253 per kWh during Off-Peak hours |

*Figure 3.3— Arizona Public Service rate schedule ET-SP; January, 2012.[137]*

This peak pricing business model doesn't apply to residential rate-payers only. It also applies to agricultural, industrial, and commercial customers.

Pacific Gas and Electric (PG&E) is transitioning agricultural customers to peak pricing. What does this mean for farmers in PG&E territory? "The average charges to operate a 250-hp (240 kW) irrigation pump may jump from $24 to $224 per hour," according to Enernoc, an energy management company.[138] Solar is already far cheaper than utility peak prices. Having to pay bills that are ten times greater than before will make more farmers consider switching to solar or wind to run their irrigation systems.

As peak pricing premiums flatten or disappear, conventional energy utilities will see their retail revenues drop and their margins squeezed.

## How Solar Disrupts Wholesale Electricity Markets

Wholesale electricity markets will also be affected dramatically by solar and wind. To see why, consider how competitive power markets work (see Figure 3.4):

1. A grid operator (also called an independent system operator) forecasts energy demands a day ahead of time. For instance, the New York ISO forecasts Q = 1,000 MW of demand from noon to 1 p.m. the following day.
2. The grid operator (ISO) asks for bids from power producers to supply the 1,000 MW quantity (Q) required to meet the forecast. Power plant operators usually bid at their marginal cost of producing the next unit of energy. Assume that a solar operator bid 200 MW at $10/MWh, a hydro-electric operator bid 300 MW at $20/MWh, a wind operator bid 300 MWh at $30/MWh, a natural gas operator bid 400 MW at $40 MWh, a nuclear operator bid 1,000 MWh at $50/MWh, and so on.
3. The grid operator starts purchasing the energy offered by the lowest bid operators until they add up to the required energy quantity (Q = 1,000 MW). In our example, the NYISO would purchase the 200 MW from the solar operator, then the 300 MW from the hydro operator, then the 300 MW from the wind operator. At this point the NYISO would have 800 MW and would need just 200 MW to complete the needed 1,000 MW. It would then take 200 MW from the natural gas operator, which would be the most expensive bid that clears the market at $40/MWh This is called the Uniform Clearing Price (UCP). The nuclear operator bid at $50/MWh would be priced out of the market.
4. The grid operator pays all suppliers the same uniform clearing price (UCP) of $40 per MWh. That is, those who bid at $20/MWh or $30/MWh still get $40/MWh for the quantity that they bid.

This method, called Uniform Clearing Price Auctions, is used by wholesale power markets in the United States.[139]

Power plant operators bid at their marginal cost, that is, at the cash cost to produce their next unit of energy (MWh). The marginal cost is mainly determined by the cost of fuel. The marginal cost of solar (and wind) is zero. Because the cost of sunshine fuel is zero, the cost of producing the next unit (MWh) of solar energy is also zero. Solar (and wind) can always clear competitive markets because they can bid at a marginal cost of zero and can therefore always clear the market and always sell for more than the marginal cost. This is not the case with operators of fossil fuel and nuclear power plants. Their marginal costs are set by the high and increasing cost of fuel.

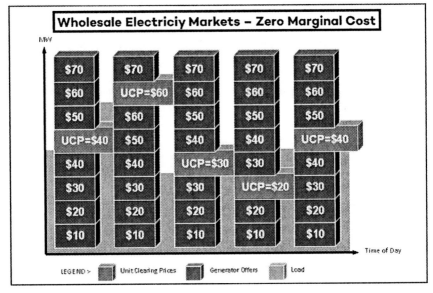

*Figure 3.4—Wholesale Markets Clearing Price auctions. (Source: NY ISO)*[140]

The clearing price of competitive electricity in wholesale markets is lower when solar and wind power plants are bidding. In 2011, each additional Gigawatt supply of solar led to an average spot price decrease of 82 Euro cents/MWh ($1.13 /MWh) in the European Energy Exchange (EEX), according to the Fraunhofer Institute for Solar Energy Systems.[141] The more solar feeds into the grid, the lower the clearing price.

The lowest marginal-cost conventional producers will set the clearing price and sell some, but not necessarily all, of the output they bid. The most expensive conventional resource-based producers (diesel, natural gas, nuclear, or coal, depending on the market) won't be able to sell an increasing percentage of their potential output. For this reason, operators of fossil fuel and nuclear power plants will see their revenues and margins squeezed at the wholesale level.

Furthermore, some of these power plants (nuclear and coal) can't produce power on demand. For technical reasons, these plants need to keep running whether or not they sell output. They literally burn cash when they can't sell their expensive electricity. Newer natural gas power plants are more flexible and can generate power on demand, which gives natural gas a technical advantage over other fossil fuels and nuclear. Newer natural gas plants don't

need to produce when their output is too expensive to clear the market.
Under the conventional business model, utilities own the generation of power, the transmission of power, and their retail business. The monopoly position that utilities enjoy has allowed them to be inefficient and still make guaranteed above-market returns on their capital. However, as electricity markets have become more competitive and as independent generators have been allowed to enter the market, the conventional utility business model has shown its inefficiencies. When there is a high penetration of solar, conventional utilities — utilities that own large fossil or nuclear generators and sell to retail end users — see their margins being squeezed on both the retail and the wholesale sides.

Both wholesale and distributed solar generation are already disrupting the conventional energy companies' century-old business model. It's not just residential users who are generating their own solar energy. Commercial customers are doing that, too.

## Exploring the Cost Advantages of Distributed Generation

A large centralized power plant has economies of scale. The plant can spread some of its costs over a larger number of production units and thereby achieve lower per-unit costs. Economies of scale may grant the centralized power plant certain unit-cost advantages over smaller distributed power generation, but a centralized generation facility requires an expensive transmission and distribution infrastructure to deliver its power to retail customers.

If you add up the costs of transmission and distribution (not to mention CEO salaries), you find that local, distributed generation has a cost advantage over centralized generation. What is this cost advantage?

The Long Island (New York) Power Authority (LIPA) performed a study to see how much it would cost to build new generation, transmission, and distribution lines to areas located east of its Canal Substation in Southampton. The study concluded it would cost LIPA more than 7 ¢/kWh in infrastructure costs to bring power to this area. Having recently suffered through Hurricane Sandy, which devastated infrastructure in Long Island and New York City, LIPA knew full well how expensive operating and maintaining transmission lines, substations, and distribution poles for the next few decades could be.

The cost of a transmission and distribution network varies from city to city, state to state, and country to country. In Europe, for example, the average cost of the network to household consumers varies substantially, from Belgium and Norway on the high end of the scale to Malta and Lithuania on the low end (see Figure 3.5). Several factors affect network costs, including voltage class, terrain type, country size, and the location of generating assets.

In the United States, the average cost of the network to household consumers in 2013 was 4.16 ¢/kWh, according to the EIA.[142]

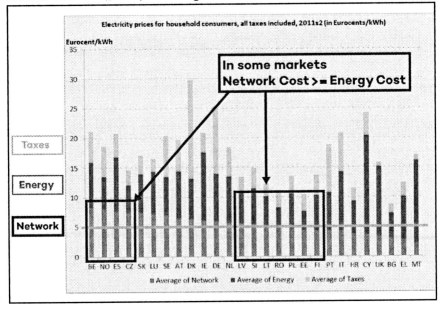

Figure 3.5—Breakdown of the cost of retail electricity in Europe. (Source: European Commission)[143]

These network cost figures are for existing transmission and distribution networks; most of these networks were built decades ago. Building a new network infrastructure is very costly. The cost of baseline technology for a transmission network ranges from $927,000 per mile for a 230 kV single circuit to $3 million per mile for a 550 kV double circuit.[144] The final cost of building a transmission line includes right of way costs and terrain multiplier costs (if you're building through mountains you can multiply the above costs by a factor of 1.7).

All of this assumes that a new distribution network can be built at all. In 2010, only 3,100 circuit miles were under construction in North America, a mere 0.7-percent increase in the 452,699 circuit miles that were already built.[145]

To avoid having to make multi-billion-dollar investments or initiate a decade-long transmission building project, LIPA decided to pay 7 ¢/kWh as an incentive for distributed local solar generation.[146] Not only would the utility save money in the long term, it would keep its capital in the bank while others invested in infrastructure.

Centralized power generation has a 7 ¢/kWh cost disadvantage compared to distributed generation. In other words, if your solar rooftop can generate energy at 10 ¢/kWh, your centralized utility power plant, to offer you a better deal, would have to generate power at 2 ¢/kWh. No resource-based centralized power producer can generate at that cost, no matter how subsidized it is. Not gas, not coal, and certainly not nuclear or diesel.

The costs of solar generation keep dropping. Soon the costs will fall below the 7 ¢/kWh (or so) cost of a transmission and distribution network. At that point, the conventional centralized utility as we know it could be wiped out.

How soon may that happen? The city of Palo Alto announced an agreement to purchase solar power for 6.9¢/kWh.[147] If distributed solar can generate solar at this cost, the utilities' days are numbered. To compete at the retail level utilities would have to sell electricity at negative prices, and that is not a sustainable business model. I don't mean that in the "environmentally sustainable" sense but in the "financially sustainable" sense.

Palo Alto's agreement is for 52 MWh of solar per year, which in its case will come from a 20MW solar power plant. This plant is much smaller than your father's nuclear power reactor (about 5 percent of the peak capacity), but it's not quite rooftop size. Power plants can be located close to where the power demand is, but not quite on site.

## Wal-Mart, IKEA, and "Big-Box" Rooftop Solar

Look at an IKEA store or distribution center and you see a massive box with lots of roof space. The economic potential of IKEA rooftops used to be so much wasted space, but not anymore. As of 2013, IKEA installed a total of 34.1 MW of solar in 39 stores in twenty American states.[148] Fully 89 percent of IKEA stores in the U.S. have solar installations.

IKEA is not the only "big-box" store that uses solar. Wal-Mart has installed 89.4 MW in 215 stores, Costco has installed 47.1 MW in 78 stores, and Kohl's has installed 44.7 MW in 147 stores. While IKEA has adopted solar in 89 percent of its stores, Wal-Mart has installed solar in just 5 percent, but Wal-Mart plans to install solar in a thousand of its 4,522 stores by the year 2020.[149] Many of these big-box stores sell solar panels in addition to using them.

Wal-Mart's average solar installation so far has been just 415kW. If Wal-Mart installed this size solar is all its stores, it would have 1.8 GW of solar capacity, a figure nearly equal to the peak power capacity of two standard nuclear reactors. As yet Wal-Mart's average solar installation is less than half of IKEA's 874 kW, but as the cost of PV continues to plummet and conventional power prices keep rising, it makes sense for Wal-Mart to build more solar installations.

Assuming Wal-Mart matches IKEA's 874 kW per installation in all its U.S. stores, it would have 3.8 GW of distributed solar capacity. Wal-Mart would then have the peak generation capacity of nearly four nuclear power plants.

Wal-Mart could also save half a billion dollars per year in energy expenses. Assuming an average energy cost of energy of 10 ¢/kWh, the local utilities would see their annual revenues go down drop by $570 million dollars for the energy that if Wal-Mart would generated energy on its own big box the rooftops of its stores.

Extend Wal-Mart's solar program to all 10,400 stores in 27 countries around the world, and the big-box retailer would have 9.18 GW of rooftop solar, matching the peak generation capacity of nine nuclear power plants.

As solar costs drop and conventional energy costs continue to rise, the value of solar to big-box stores rises, which is a powerful financial incentive for the stores to generate even more solar energy.

Utilities stand to lose billions in yearly revenues if big-box stores generate more of their energy onsite using solar. Lower demand from commercial customers, especially during "peak pricing" periods, will flatten the utilities' peak premiums, lowering their revenues and squeezing their margins.

IKEA wants to generate all its energy needs using solar and wind by 2020.[150] Wal-Mart has set itself the goal of generating 100 percent of its energy from clean sources.[151] As more large organizations become net-zero organizations, conventional utilities will find themselves losing their biggest customers altogether.

A vicious cycle of lower revenues, lower margins, lower capacity utilization, and lower returns on investment will set in. Conventional utility capital costs will rise, making their energy even more expensive, and this, in turn, will feed into the vicious cycle.

Wal-Mart, IKEA, and other big-box stores becoming net-zero organizations will be bad enough for utilities, but it may not be the utilities' worst-case scenario.

The next step in the disruption process happens when these companies go from generating their own energy to supplying energy.

Utilities are used to cushy monopoly rents. Competition is not in their DNA. As big-box stores generate surplus energy, utilities may find themselves competing against their former customers. Utility executives don't want to compete with small Silicon Valley startups like Sungevity, SolarCity, and SunRun. Utilities can hold them back temporarily by making use of the public regulatory system, which they built and have fed for over a century. However, the thought of a powerful, wealthy, tech-savvy and hyper-competitive company like Wal-Mart entering the energy market by selling cheap solar energy will certainly make utility managers double-check their retirement accounts.

In a distributed-energy-generation world, big-box stores entering the energy business is bound to happen. Remember, the marginal cost of solar energy is zero. Wal-Mart and other big-box commercial generators would always clear prices in a competitive wholesale electricity market.

Wal-Mart and other big-box retailers may even decide to sell directly to consumers. According to a recent report by Accenture, 59 percent of energy consumers would consider buying electricity directly from a retailer like BestBuy, Tesco, or Carrefour.[152] The report also found that 47 percent of energy consumers would consider buying from an online-only company. Can you see Amazon.com in the power business?

Conventional utilities would use their best weapon to fight competition: the regulatory system. But disruption can't be stopped, only postponed a bit.

Big-box stores are not the only ones to realize the cash value of rooftop solar plants. Industrial companies such as VW, Nissan, and Apple have announced fairly large onsite solar projects. The German automotive giant Volkswagen recently announced the completion of an 11MW rooftop solar plant in Spain.[153]

Apple, the world's largest company by market capitalization, has a goal of being a net-zero company, 100-percent powered by clean energy. In 2012 Apple built a 20 MW ground-based solar power plant on land surrounding its data center in Maiden, North Carolina. According to the company, its plant is the largest ever onsite end-user owned solar PV plant in the world. The company is also building a second 20 MW solar PV plant within a mile of the data center. This plant is expected to be operational by late 2013.[154] According to Apple, its new Reno, Nevada, data center will be 100-percent powered by solar and geothermal energy.

## Real Estate Managers Discover Solar

Prologis is a leading owner, operator, and industrial real estate management company with properties around the globe. In the Americas alone, Prologis manages $46 billion in assets, including approximately 563 million square feet (52.3 million m$^2$) of logistics and distribution space.[155] That's a lot of rooftop space that doesn't generate any income. Can a real estate management company change this equation by adding solar to its portfolio of income-generating assets? It sure can.

In June 2011, Prologis announced a partnership with NRG Solar and the Bank of America to build 753 MW of solar on Prologis-owned warehouse rooftops.[156] This would generate enough solar to power a hundred thousand homes. Bank of America will lend $1.4 billion to the project. Prologis has already built 34 installations totaling 79.6 MW.[157]

Other real estate developers and managers have seen the light. Hartz Lights Industries built 17 solar installations totaling 19.2 MW. Kimco built three MW on six sites.

Prologis and Hartz Lights Industries manage hundreds of millions of square feet of industrial facilities. As they prove that distributed solar can be a money earner, thousands of warehouse owners around the nation will take notice.

## The Robot Thermostat

Solar is not the only technology disrupting the power utilities. Robots and other Silicon Valley technologies are starting to cut into the utilities' business profits.

In 2007, Yoky Matsuoka received a MacArthur Fellowship "genius" award. The MacArthur Foundation said her work "transform(ed) our understanding of how the central nervous system coordinates musculoskeletal action and of how robotic technology can enhance the mobility of people with manipulation disabilities."[158] When Matsuoka received the award, she was an associate professor of computer science and engineering at the University of Washington, where she directed the University's Neurobotics Laboratory and its Center for Sensorimotor Neural Engineering.

Matsuoka earned a Ph.D. in electrical engineering and computer science from the Massachusetts Institute of Technology. She practically created the field of neurobotics (the combination of neurology and robotics). Her MacArthur grant awarded her more than $200,000 to spend as she saw fit over the following four years. What did Dr. Matsuoka do? She returned to Silicon Valley where, after a brief stint at Google, she joined a small startup called NEST

Laboratories as its vice president of technology. NEST was started by two of the original Apple iPod designers who wanted to do in the field of energy what Apple had done in the music field.

In 2012 NEST launched the $299 NEST Learning Thermostat. This was no ordinary programmable thermostat. It was a sensor-based, Internet-enabled, artificial-intelligence-based computer that continuously scanned the temperature of a home and learned about its inhabitants' preferences.

Heating and cooling account for about 56 percent of energy usage in a typical U.S. home, according to the U.S. Department of Energy.[159] Electricity bills can be painful when air conditioners are at full throttle in the summer. Electricity consumption goes up by more than 40 percent in the late afternoon during the hottest days of the summer.[160] When most everyone turns up their air conditioners, the cost of wholesale electricity goes up by more than 100 percent. As a result, homes consume more electricity when it is most expensive to use electricity. This is why energy bills climb into the hundreds of dollars so quickly in the summertime.

In fact, in many markets during the summer, "peak" electricity prices can be many times higher than "minimum" or "average" prices. During a heat wave in August 2011 in Texas, prices peaked around $6/kWh, up to ten times the "normal" on-peak pricing (see Figure 3.6). It's ironic that the price rose to $6/kWh in Texas, over fifty times the cost of solar, on a sunny day when solar generates the most electricity.

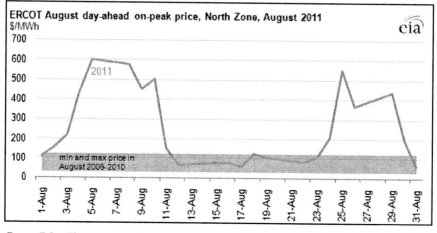

Figure 3.6—Electricity peak prices in Texas, August 2011. (Source: EIA)[161]

Super-hot days are not going away. In 2012 more than 27,000 daily high-temperature records were broken in the U.S.[162] The June 2011–June 2012 year was the hottest year on record for the contiguous United States, according to the National Oceanic and Atmospheric Administration.[163]

The smart use of air conditioning can lower energy usage substantially. Programmable thermostats have been sold for decades but they were difficult to use because it was hard to program them. Customers usually stopped using the thermostats after a few weeks.

Yoky Matsuoka and NEST set out to change that. Using artificial intelligence, the NEST thermostat tracks a user's preferences and comfort level. The user sets the temperature with one simple dial (think iPod). Sensors in the thermostat detect when the user is home and the thermostat changes the temperature accordingly based on data gathered about the user. To minimize energy usage, the thermostat lowers the temperature when the user leaves for work. An app that runs on iPhones and Android smartphones makes it possible for the user to tell the thermostat to turn the heater or air conditioner on or off. In fact, a user who comes home from work at 5:45 p.m., for example, can program the thermostat to pre-cool the home by 5:45.

During the first year of operation, the NEST thermostat already saved its users more than 50 percent on their cooling bills.

The NEST thermostat was just the first iteration of the product. Remember, this thermostat is a computer like the iPhone is a computer. As NEST develops new features for its thermostat, users can download these features to their thermostats via wireless Internet.

All the technologies that the thermostat is built upon are improving exponentially: sensors, machine learning, wireless communications, big data, and distributed computing. Sensors are getting cheaper, faster, better, smaller, more connected, and more energy efficient. Sensors are embedded in smartphones, tablets, and wearable devices. A Samsung Galaxy S4 phone, for example, has sensors for detecting motion, light, temperature, humidity, location, and more (see Figure 3.7).

One of the world's leading suppliers of sensors is Silicon Valley-based Invensense, Inc. Speaking to my class at Stanford University, Steve Nasiri, the former CEO and founder of Ivensense, said that a motion sensor that used to cost $25 a decade ago costs barely $2. Today's sensors are also a hundred times smaller and ten times more energy efficient. Sensors will probably cost mere cents in another decade.

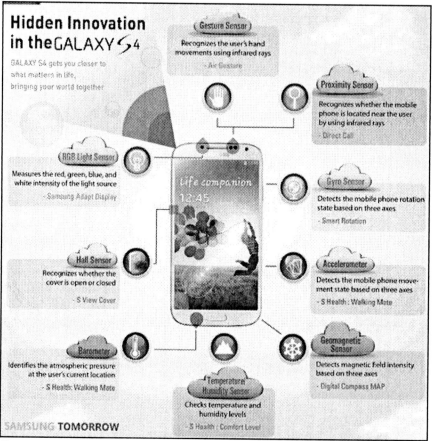

*Figure 3.7—Sensors in a Samsung S4 Smartphone. (Source: Samsung)*[164]

NEST is turning energy efficiency into an inexpensive, quick, and easy-to-adopt technology product. Energy efficiency has traditionally been about changing the configuration of the home: new windows, new wall insulations, and LED lights. These things are important, but so is a smart thermostat. By spending $299 and five minutes to install it in your home, you can save up to 50 percent on your heating bills. If millions of people do it, they will save hundreds of millions.

Power utilities take note: Electricity usage may go down quickly and significantly because of smart information technology devices like the NEST Learning Thermostat. Like rooftop solar, smart energy management devices will contribute to the vicious cycle of utility disruption.

As more customers manage their energy usage with products that employ artificial intelligence, total demand for expensive peak energy (kWh) will go down. Utilities will have to divide their sunk costs by a smaller denominator, which will make electricity bills go up. Utilities will add fees and raise energy prices. Consequently, more users will switch to solar and to smart energy devices, which will lower demand from utilities even further. The cost of capital to utilities will go up as their revenues decrease, making their services even more expensive. Once this vicious cycle starts it's hard to stop it.

Furthermore, companies like NEST are just getting started. Recently, at a Robotics conference in San Jose, California, NEST's Yoky Matsuoka announced the company's new device: the Protect, a smoke alarm and carbon monoxide detector.

A smoke alarm is not normally the kind of product that utilities worry about, but in the case of Protect, utilities should be alarmed because the device comes with numerous sensors that have something to do with detecting energy flow in the home. Here's the list of the sensors in the smoke alarm:
- Photoelectric smoke sensor
- Carbon monoxide sensor
- Heat sensor
- Ambient light sensor
- Humidity sensor

Every second of every minute of every day, Protect will collect a multi-dimensional array of energy information for each home in which it is installed. Protect will collect more energy data than anyone or any device has ever collected before. NEST will have more information about home energy usage than conventional utilities could ever dream of.

Certainly, the purpose of the data is to protect the home from smoke and fire. But analyze the data intelligently and the possibilities for developing new products and services for the home are boundless.

In 2014, Google acquired NEST for $3.2 billion. Imagine the power of Google's vast computing resources and data infrastructure behind NEST energy-management devices.

## How Big Data Increases Returns on Clean Energy

Climate Corp., a San Francisco-based company, is a good example of how Silicon Valley technologies can unlock the value of big data. The company was founded in 2009 by two former Google employees. Using weather data from the U.S. government, Climate Corp. creates products to improve agricultural crop yields. The company combined thirty years of weather data, sixty years of crop yield data, and 14 terabytes of soil data for uses such as researching and pricing crop insurance.[165] In October 2013, Climate Corp. was acquired by agriculture giant Monsanto for $930 million in cash.

The data that Climate Corp. used was freely available from the U.S. government. Climate Corp. created information technology for unlocking the value of that data. Imagine what a company like NEST can do along similar lines. By combining data about weather and energy from the U.S. government with the private data it obtains from its billions of sensors in tens of millions of homes, NEST can unlock the value of data as never before.

To a utility, a home is like a black box; pour energy in and cash comes out. Utilities really don't have much information about their customers.

By contrast, NEST is building a knowledge base to design products that build on one another — and disrupt industries, including energy, in the process. Like an iPhone that is connected to an iPad, the NEST Learning Thermostat and the Protect smoke alarm will be connected to one another; they will be able to communicate and learn from one another. For example, according to Yoky Matsuoka, 40 percent of fires start in the furnace. Knowing from the thermostat device when the furnace is heating up can help the alarm device determine whether the smoke it senses is likely caused by a genuine fire, not by overcooked beef on the barbeque grill.

NEST is a prime example of Silicon Valley's economics of increasing returns. For a homeowner, purchasing the smoke alarm increases the value of their learning thermostat — and vice-versa. Similarly, if your neighbor purchases a NEST smoke alarm it decreases the likelihood of a fire in your home (and theirs, of course). As more neighbors purchase learning thermostats, NEST will be able to analyze whole neighborhoods and improve their products with the data it collects. Everyone who purchased a learning thermostat will benefit when they download the new and improved software version.

Today's thermostat is collecting the data and building the customer insights to develop tomorrow's energy-management platform and, who knows, maybe a future energy-trading platform.

## The Zero Disruption:

## The Exploratorium Science Museum

Science is disrupting the power business in many ways. In San Francisco, the Exploratorium science museum is showing the way to zero.

Recently, the Exploratorium opened the doors to its brand-new facilities on Piers 15 and 17 along San Francisco's Embarcadero. The New York Times called the new museum "the most important science museum to have opened since the mid-20th century." [166]

Previously housed next to the Palace of Fine Arts in the Marina District, the Exploratorium tripled its display space to 330,000 square feet, built a 400-person theater, and moved its staff back inside the same building as the museum (previously the staff had offices in a separate building). The museum expected to double its attendance from a half million to one million visitors per year.

For all that growth, one thing went down to zero: its power bill. The Exploratorium is designed to be a net-zero energy campus, powered 100-percent by solar panels on its rooftop. The museum has built a 1.4 MW solar array that it expects will generate 2.1 GWh of energy in its first year of operation. The museum expects solar power to cover all its annual power needs. This energy consumption would otherwise cost about $300,000 per year.[167]

Solar panels generate all the energy, but this isn't the only reason why the Exploratorium is a leader in 21st-century energy architecture. The Exploratorium was designed with energy-efficiency features. The building consumes less than 50 percent of the energy a similar building would use, according to Chuck Mignacco, the building operations manager.

Built on a pier on the waterfront, the museum uses an innovative water and energy management system called the "bay water heating and cooling system." Eight solar-powered motors pump 73,800 gallons of bay water; the water is then filtered, sterilized, and circulated throughout building (see Figure 3.8). The filtered water is used for heating and cooling, depending on the season. Every office, every exhibit, and every laboratory is heated or cooled using bay water. Even the data center on the second floor is cooled using bay water. This ingenious HVAC system allows the museum to use 80-percent less energy for cooling and 77-percent less energy for heating than industry-standard buildings.

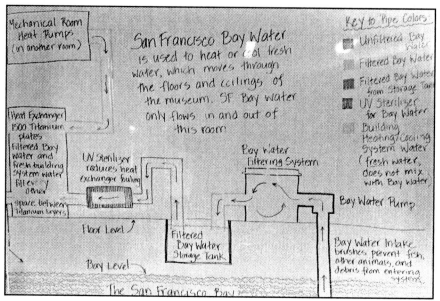

*Figure 3.8 — The San Francisco Exploratorium's bay water heating and cooling system. (Photo: Tony Seba)*

The conventional way of cooling and heating the building would be to use cooling towers, which are both inefficient and visually obtrusive, and to use natural gas to power the system.[168] But the Exploratorium is not a conventional building. Inside its buildings everything is powered with solar electricity — the LED lights, water pumps (for heating and cooling), ventilation fans, computers, and laboratory equipment. The museum's carbon footprint would be zero except for a one-inch natural gas pipe that is used in the kitchen.

## How Distributed Design Helps Clean Energy

On July 24, 2013, Mike Sami and I were in his office in Millbrae, California, analyzing a 400 MW wind power plant that we were developing for the Republic of Georgia. A wind resource assessment engineer in Spain sent the designs via email the previous day. Looking at the whole plant on Google Earth, we could click each individual turbine to see a pop-up information box with key metrics: height, coordinates, wind speed, and so on. Mike noticed that some 3 MW turbines were on ridges. He asked if I thought they should be moved, and, if so, should they be placed closer to a ridge that already had a half-dozen turbines.

You have to strike the right balance when designing a wind power plant. You can't place turbines too close together because the turbines affect one another's wind resource and production. However, because you need to build transmission lines and roads to each and every turbine, you benefit from lining them up on the same ridge. We run computer simulations to find the configuration that optimizes energy generation while minimizing costs. This is a necessary step in the power plant design process.

At the end of the afternoon, we emailed our feedback to the engineer in Spain and called it a day. The following morning, I woke up at my usual 4 a.m. Not much later, I received a reply email from Madrid with three proposed wind turbine configurations.

It occurred to me that designing a wind or solar energy power plant is very much like an information technology programming project. In both cases, highly skilled individuals collaborate over the Internet using open data, big data analysis, open source technology, and knowledge that others had previously created and shared with the world.

The wind data for the Republic of Georgia plant was publicly available from NASA and the U.S. Department of Energy. Google had built Google Earth and the tools to create maps and the designs to take advantage of these maps. The wind simulation software was originally an open source program; dozens, maybe thousands of programmers and wind energy engineers had given millions of hours to build this tool and make it available to others. Anyone who wanted to improve this software could do so.

The simulation exercise might even be run over a network of computers in several countries to take advantage of downtime in many people's computer usage. I imagine our Spanish wind assessment engineers used computers in their American office that were idle after the American team went home. Or they may have used a cloud service like Amazon's Web Services (also a standards-based, openly available platform). All the elements that had contributed to the growth of the Internet itself were there.

Electricity 2.0 is both an information technology and an energy infrastructure. As such, it is governed by information economics. Like computing, solar and wind are based on economies of increasing returns.

## Meet Rachel Rhodes, Remote Solar Disrupter

I recently paid a visit to Sungevity, one of the fastest growing solar installers in Silicon Valley. Sungevity designs, finances, installs and operates solar power plants on residential rooftops.

Sungevity doesn't make solar panels, inverters, or any other system hardware. The company doesn't own trucks or have any hardware inventory. They don't even climb their customers' rooftops or install solar panels. Ever. Not to measure them, not to install the panels, not to maintain them.

Sungevity is a software and finance company. From the moment a prospective solar customer signs up on the Sungevity website (or on its affiliate network website), everything is done from the company's office in Oakland.

To design the residential rooftop solar installation, Sungevity has a team of remote designers. I talked to Rachel Rhodes, who showed me how she designs a solar installation for a rooftop hundreds of miles away.

Rachel Rhodes graduated from Tufts University with a degree in international relations and environmental science. She has been at Sungevity for a year and a half. She first opens a window on her computer screen where she gets the prospective customer's street address. Someone at the company has already spent thirty or forty minutes on the phone talking to the customer, analyzing the customer's energy consumption, explaining the financing options, and so on. Rhodes then looks at a bird's-eye view of the house on Google Earth. Roofs come in many shapes and angles. Rhodes has many ways to place the panels on the rooftop to maximize energy production and minimize costs.

Every jurisdiction in the U.S. has different building codes and design requirements. For instance, some cities require a three-foot setback, which diminishes the potential surface available for solar panels. Rachel Rhodes looks at the trees and the potential shading. On a second 20-inch screen, she has a slightly different bird's-eye view of the rooftop she is working on. This second view allows her to look at the rooftop at an angle so she can study the shading.

She then starts placing the virtual panels on the rooftop. The customer has already told Sungevity how much electricity he or she consumes. Rhodes takes this data into account when designing the configuration and size of the panels. It's as if she were playing a solar version of a computer game like Tetris.

Less than ten minutes later, she decides on the right configuration. Company co-founder Danny Kennedy told me designing remotely is faster and more accurate than having "boots on the roof" and taking measurements there. Sungevity's solar disruption is producing "faster, cheaper, and better" products and services than the competition can produce.

Based on Rhodes' design, the customer gets an "iQuote" — a plan for getting solar power for the next twenty-five years. If the customer signs the agreement, Sungevity software automatically starts the regulatory paperwork process, contacts a local independent company-certified installer, and initiates the logistics for the delivery and installation of the solar panels.

The solar installation itself takes only a few hours, depending on the home and design configuration.

However, the regulatory paperwork, which is controlled by local jurisdictions and the utilities, is the bottleneck that may postpone the installation for weeks or months.

Rhodes has used the software to design solar rooftops outside the United States, in Australia and Holland, from her cubicle in Oakland. Was there any difference with Australia? She said you have to remember that in Australia the panels pitch north.

The solar industry involves highly skilled individuals collaborating over the Internet who build using open data, big data analysis, and open-source technology. They take advantage of knowledge that others previously created and shared with the world.

Google built Google Earth and the tools to create maps and the designs to take advantage of it. The solar insolation data was based on publicly available information from NASA and the U.S. Department of Energy. Individuals and companies around the world are contributing content, technology, and skills to improve these tools. Sungevity takes this technology mash-up and adds its unique technology skills and intellectual property. Here you have the Silicon Valley ethos in a nutshell.

Silicon Valley bits have blended with solar electrons to create an open, scalable Internet-based infrastructure. The economics of this bits-and-electrons Internet infrastructure is based on increasing returns against which the extraction-based, command-and-control, atom-based energy industry can simply not compete.

I walked away from my meeting with Rachel Rhodes thinking that it's not hard to imagine this infrastructure building millions or tens of millions of solar rooftops and disrupting utilities in the blink of an eye.

## Utility Lobbies of the World:
## Unite and Raise Prices!

The Edison Electric Institute, an association and lobbying organization for investor-owned utilities (IOU) in the United States, recently published a report titled "Disruptive Challenges: Financial Implications and Strategic Responses to a Changing Retail Electric Business." The report highlighted the disruptive threat of solar and distributed energy resources (DER) to its member utilities. The Edison Electric Institute tells its members that it isn't necessary for utilities to change their business model. Instead, they should go to the public utility commission and ask for more money from ratepayers:

> So, while the telecom example is a tale of responding to the threat of obsolescence, the near-term challenge to the electric sector is providing the proper tariff design to allow for equitable recovery of revenue requirements to address the pace of non-economic sector disruption.[169]

Translation: "Don't worry, just raise prices!" Some of the "immediate" and "long-term" actions it recommends include:
- A monthly customer service charge.
- Charging ratepayers a fee to help the utility invest in new equipment.
- Charge a fee to customers for leaving the utility.

In an era of increasing customer choice and distributed clean energy at lower prices than conventional sources, the electric utilities are being told to raise the barriers to consumption. The Edison Electric Institute promotes the "head in the sand" strategy. Just wait for your Kodak moment! It will come soon enough!

In the meantime, utilities in Europe, where clean energy adoption is leading the rest of the world, are already feeling the pain of disruption. Since its peak in 2008, the top twenty European electric utilities have lost half their market valuation, dropping from a trillion Euros ($1.3 trillion) to half a trillion Euros ($650 billion) in market capitalization.[170]

German-based utility giant E.ON is an example of the disruption of utilities in Europe (see Figure 3.9). As of October 2013, its American Depositary Receipt (ADR) stock had dropped by more than half to below 15 from its highs near 30 in 2009 and 2010.

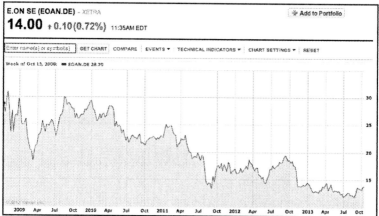

*Figure 3.9—E.ON stock chart – 2009-2013. (Image Source: Yahoo!)*

And Edison Electric's debt nightmare is already happening. In 2008, all of the top ten European electric utilities had a credit rating of A or better. By 2013 only five had the A rating.

A drop in a utility's stock price means that the equity value drops, which implies that capital investment will drop. The deterioration in credit ratings means that the interest rates that a utility has to pay on the power plant mortgage goes up. This combination of less equity and higher interest payments delivers a one-two punch increase in the cost of capital. The higher cost of capital means that fewer projects will be NPV (net-present-value) positive so fewer conventional power plants will get built. It also means that those plants that do get built will produce more expensive power because their interest rate payments are higher.

This expensive power cannot compete with ever-decreasing cost of solar (and wind). Conventional utilities are already falling into a vicious cycle. As they become less competitive, they lose customers to solar and find it harder to raise capital at low rates. They become even less competitive.

## The Next Disruptive Wave:
## Distributed Electricity Storage

Conventional power utilities will soon be hit by several disruptive waves. Each wave will carve a large hole in the utilities' century-old business model. Networks of sensors, machine learning, and connected devices allow for the distributed generation of power and customer-centric energy management. All this technology will hit utilities were it hurts most by
- Lowering the clearing price of wholesale competitive markets
- Flattening the peak premium pricing in retail markets
- Lowering demand because of increased end-user self-generation
- A critical mass of customers going net-zero

This one-two punch of decreasing revenues and dramatically lower earnings would scare most industrialized organizations, but not power utilities. Utilities are so deeply entrenched in the regulatory process, they don't have to reorganize to meet advances in technology, changes in consumer preferences, and market changes.

The conventional landline telephone companies did not die. When the world adopted mobile phones in the early 1990s, most people kept their landlines as a backup. But as the quality of cell phones improved and users got comfortable with them, they unplugged their landline phones. Users who grew up with cell phones never made the acquaintance of the landline. Developing countries that had no landline infrastructure to speak of just leapfrogged to a mobile communications infrastructure.

During the current distributed solar disruption wave, homes and businesses will use the utilities as a battery backup — an expensive but necessary backup. Power utilities have existed for a century without having to compete for customers. This state of monopoly bliss is nearly over. Now the utility has to share the customer (or "rate payer") with two technology-based, exponentially improving companies: the PV provider and the energy-management provider.

The power utilities will soon be experiencing a vicious cycle of lower demand and rising costs while the market provides lower costs and a higher quality of service.

The next wave of disruption will occur when electricity storage is cheap enough for users to store some of their daily production or usage. A convergence of technologies is lowering the cost of solar power and increasing the quality exponentially:
- The cost of batteries is dropping and the quality of batteries is improving.
- Intelligent energy management devices are making energy usage more efficient. These devices are enabled by rising quality and lower costs in machine learning software, sensors, and communications think NEST).
- Solar PV costs are dropping.
- Utility costs are rising.

## The Next Disruptive Wave:
## On-Site Electricity Storage

The second wave of disruption will happen as users start adopting energy storage and intelligent energy-management devices.

Electricity storage companies have learned business model innovation from solar providers. Recently a Silicon Valley company called Stem started offering "storage as a service." Similar to solar third-party providers, Stem offers customers a "no money down" agreement for its energy storage and management services. The company owns, finances, installs, and manages the storage technology in exchange for a ten-year commitment from the customer.[171]

Stem is not just offering electricity. The company offers to lower electricity costs by 10 to 50 percent. How? The old-fashioned way: By buying electricity when prices are low, storing it, and using it when prices are high.

The Silicon Valley virtuous cycle of talent and money is at play here. Jigar Shah, who created the concept of "solar as a service" in 2008, sold his company SunEdison to MEMC for $200 million at the end of 2009.[172] He then created the Carbon War Room, an NGO think-tank, and became the lead investor in a venture fund called Clean Feet Investors (CFI).

Stem recently announced a $5 million round of investment led by CFI. They didn't just learn from the master; they got his money too!

How soon will the disruptive wave enabled by on-site electricity storage come? I crunched some numbers. In Chapter 4 I write about the evolution of the cost of Lithium-Ion batteries for electric vehicles. Li-on can also be used for electricity storage in homes and businesses. Many other technologies are being developed for grid storage but I use Li-on to illustrate the evolution of costs and their effects on electricity markets (see Table 3.1).

| Purchase Cost of Battery Storage System ($/kWh) -> | | | $600 | $500 | $300 | $200 | $100 | $50 |
|---|---|---|---|---|---|---|---|---|
| | Hours | kWh | Monthly Cost of Storage | | | | | |
| Demand Response | 1 | 1.25 | $4.6 | $3.8 | $2.3 | $1.5 | $0.8 | $0.4 |
| Avoid peak, buy low & Shift Usage | 4 | 5 | $18.4 | $15.3 | $9.2 | $6.1 | $3.1 | $1.5 |
| Store all Solar self-generation | 8 | 10 | $36.8 | $30.7 | $18.4 | $12.3 | $6.1 | $3.1 |
| Self-sufficiency | 16 | 20 | $73.6 | $61.3 | $36.8 | $24.5 | $12.3 | $6.1 |
| Off-grid | 24 | 30 | $110.4 | $92.0 | $55.2 | $36.8 | $18.4 | $9.2 |

Table 3.1— Capital costs of Li-on battery storage vs. monthly levelized cost of storage.

In the future, electricity consumers will store some of the electricity they consume at their residential, commercial, or industrial sites. Storing electricity at the point of use provides consumers with several benefits. At a minimum, someone who stores the equivalent of a few hours of daily demand can purchase electricity (from a solar rooftop or from the grid) when electricity costs are low and use it when electricity costs are high. Storing four hours of power, for example, would help consumers avoid the high cost of power during peak usage cycles. This can make a difference of hundreds of dollars per month during the summer months when usage is at its peak.

The average residential user in the U.S. consumes about 903 kWh of electricity, a little over 30 kWh per day.[173] In order to store four hours of usage, a homeowner would need to purchase a 5 kWh storage system. Today, Lithium Ion batteries plus the electronics device to manage them might cost about $600 per kWh, making the capital cost for this system around $3,000. Assume that a hypothetical user finances the $3,000 at 4 percent cost of capital over twenty years (mortgage rates). The monthly payment for this system (levelized cost of storage) would be around $18. Depending on the user's monthly electricity bill, such a system could pay for itself in a few hot summer months. After that, it's pure savings for the user; for the utilities, it's a loss of revenue during peak power usage.

Now assume that same hypothetical user decided to store a third of his daily usage (10 kWh). The monthly cost would be $36.80. This user would get the benefit of buying energy at the lowest price plus the benefit of storing the excess rooftop solar energy that he doesn't use during the day for evening usage. The user could even make money when the utility runs demand-response programs (he would get paid for not consuming grid electricity during peak periods).

For a user to substantially generate and consume only his own energy, he would need to generate rooftop solar and store 20kWh. The capital cost for this system at $600 per kWh would be $12,000, or a monthly storage cost of $110.40.

But the costs of Li-ion technology are plunging. The current consensus is that Li-on will be on the order of $200 to $250 per kWh by 2020. At $250/kWh, a user could essentially not pay for peak prices for $7.70 per month. A user could, for about $15.30 per month, have eight hours of storage to shift solar generation from day to evening, not pay for peak prices, and participate in demand-response programs.

As utilities follow the guidelines set by Edison Electric Institute and raise electricity rates and fees, as they balk at buying excess solar generation from end users, they make it easier for energy consumers to go solar and invest in energy storage.

Utah's Rocky Mountain Power (RMP) recently asked its public utility commission to approve a minimum monthly bill of $15, a customer service fee of $8 per month, and a monthly solar fee of $4.25. That's a minimum monthly bill of $27.25 plus actual usage.[174] Fees like this make on-site storage more financially viable. At $600 per kWh it would cost a user $18 per month to have four hours of on-site storage to store excess solar generation. By 2020 this number is expected to fall to $7.70 per month. Not only will solar generate cheaper energy than utilities, it will make more financial sense for users to store energy on-site rather than share it with the grid.

By 2025 it will cost just about $12.30 per month to have 20 kWh of onsite storage. For the average American residential user 20 kWh represents two thirds of daily consumption. Fifty or sixty million American homeowners will essentially be able to generate all their energy using solar; they will be able to store what they don't use at the moment of generation for the rest of the day and night. For just $18.30 per month they will store enough energy to essentially get off the grid.

By inventing new fees, raising existing fees, and raising energy prices, the utilities are increasing their short-term cash flow at the expense of their very survival. By increasing the price of their service at a time when the cost of distributed solar generation is dropping dramatically and the cost of on-site storage is becoming competitive, utilities are actually helping to accelerate the adoption of these technologies. Utilities are helping their own Kodak moment to arrive faster.

## Kodak: An Example for Electric Utilities

The century-old electric utility business model is fundamentally obsolete. The ever- increasing penetration and ever-lower costs of distributed solar energy have already started to affect the centralized monopoly big-energy business model.

It's tempting to say that power utilities are like Kodak watching digital photography erode its business. Kodak is a classic market disruption story, but don't confuse stories of disruption with clueless management.

Kodak essentially invented digital cameras. The company invested billions in digital imaging technology over decades. For instance, in 1986 Kodak invented the world's first megapixel sensor (1.4 million pixels) that could produce a 5x7-inch digital photo-quality print.[175] The following year, Kodak released seven products for recording, storing, manipulating, transmitting, and printing electronic still video images.

In 1991 Kodak released the first professional digital camera system for photojournalists. It was a Nikon F-3 camera equipped by Kodak with a 1.3-megapixel sensor. By 2001, Kodak had invested $5 billion in R&D. It held more than a thousand patents on digital-imaging technology.

In 2001, Kodak's CEO Patricia Russo said, "Kodak must move beyond photography to capture a major slice of what it forecasts as a $225 billion market for so-called info-imaging, a world where data, audio and images converge. Kodak is poised to dominate this arena."

Kodak had the technology, knew the market, and had a brand that was synonymous with photography. Yet Kodak was unable to react to market changes. Why? Because it had the old business model DNA. It believed that buyers who bought cameras were hooked on film. They had to keep buying film for the rest of their lives. Every time a photo was taken the cash register would go "ka-ching." Film was a cash cow.

The digital business model was different. In this model, after the supplier sells the camera, they're basically done. From the consumer perspective, the only meaningful cost is the digital camera itself. The marginal cost of generating, processing, transmitting, and consuming pictures is basically zero.

Even when Kodak wanted to adapt to digital it couldn't do so. Its business model DNA would not allow it. At one point it even developed a "hybrid" product that combined digital and film — an "all-of-the-above" strategy, if you will. Kodak had to preserve the old model at all costs. It turned out to be a "none-of-the-above" outcome.

Once the market transition from film to digital photography started in earnest, it was swift. It took less than a decade for Kodak to go from industry leader to filing for bankruptcy protection.

Substitute "digital cameras" and "photography" for "solar" and "electricity" and you begin to understand the pattern. Many utility executives know the disruption is coming. They're just addicted to the cash flow and can't let go of their conventional business model. The more they want to increase the short-term cash flow, the more they enable the disruption.

## The Twenty-Year Itch: How Solar Outperforms Conventional Generation

On January 24, 2004, NASA's Opportunity Rover landed on Mars for what was expected to be a three-month exploration mission (see Figure 3.10).[176] NASA expected the Rover to travel about one kilometer (0.6 miles) before the solar panels that powered it were covered with dust and could no longer generate the energy needed to power the vehicle and its laboratory instruments.[177] Instead of three months and one kilometer (0.6 miles), the solar panels powered the Rover for ten years and 38.7 kilometers (24 miles). Along the way, the Rover has taken 170,000 pictures and uploaded them to Earth, which is on average 255 kilometers from Mars.[178] The power of solar, it seems, has always been underestimated.

*Figure 3.10—The Opportunity Mars Rover. (Photo Source: NASA)*[179]

By contrast to solar, fossil fuels and nukes are so toxic and corrosive, thermal power plants are damaged beyond repair after forty years. Often during their four decades of use, they have to shut down for months or years at a time for repairs and maintenance. As the existing fleet of nuclear and coal power plants gets older and more inefficient they also get more expensive to operate and maintain; they increasingly become less competitive. A report by investment bank CreditSuisse points out that the number of nuclear plant outage days has increased significantly, necessitating higher capital costs for repairs and upgrades (see Chapter 6.)

The solar panels that drive the Opportunity Rover were expected to last three months because conditions on Mars are brutal: extreme temperatures, radiation, dust storms, and so on. And yet, ten years on, the Rover's solar panels keep generating energy. The solar panels have lasted forty times longer than originally expected. And they're still going.

Back on earth, solar PV production on the rooftop doesn't fall off a cliff after twenty years. While solar panels do fall in efficiency each year, the drop is so small it's barely noticeable on a year-to-year basis. Estimates assume a drop of 0.5 percent each year; after 20 years a solar installation is expected to produce about 90 percent of what it produced its first year. At that point, after 20 years, the solar power plant will have paid off the mortgage. After 20 years, those power plants essentially generate power for free. For life. Plant operators have to change the inverters every ten years or so, but otherwise solar power plants are cash machines for many decades to come.

Germany, which started its solar program around 2000, will find itself with Giga-watts of solar installations generating power for free in 2020. Solar power plants that went online in 2010 will start producing free power in 2030, and so on. What happens after 2040 or 2050 when most of the solar power plants in Germany have a zero total cost of producing energy? The country will have the cheapest power prices in the world!

## The Empire Strikes Back:
## David vs Goliath in California

California has been called a "chaotic" and "messy" direct democracy.[180] The ballot includes referendums by which voters can reject acts of the legislature and initiatives by which voters can write their own laws. The political system in California, as in many other places, has been hijacked by lobbyists and special interest groups who pursue their interests at the ballot box by claiming their initiatives are in the public interest.

The June 2010 ballot included Proposition 16, an initiative called the "New Two-Thirds Requirement for Local Public Electricity Providers Act," according to BallotPedia.[181]

> If Proposition 16 had been approved by voters, it would have henceforward taken a two-thirds vote of the electorate before a public agency could enter the retail power business. This would have made it more difficult than it is currently for local entities to form either municipal utilities, or community wide clean electricity districts called Community Choice Aggregators (CCAs).

Pacific Gas and Electric (PG&E) wanted to stop the trend toward distributed, participatory, local energy generation. The board of PG&E, the largest utility in America, approved a war chest of $35 million to support the initiative.

The main targets of this initiative — local governments, irrigation districts, and municipal utilities — were prohibited by law from spending any money to fight it, according to former California Energy Commissioner John Geesman.[182]

However, there was a groundswell of opposition to Proposition 16 and PG&E. More than 38 newspapers around the state published editorials opposing what they considered a "power grab." Forty cities and ten chambers of commerce publicly came out against it. In campaign literature, PG&E called its coalition "Taxpayers for the Right to Vote," a rouse intended to give the impression that Proposition 16 was about people's "right to vote."[183]

PG&E ended up spending $46.1 million on Proposition 16 against just $100,000 that opponents scrapped together.[184] Despite this 461-to-1 spending advantage, California voters defeated Proposition 16 by a 53 to 47 percentage of the vote.

California's Proposition 16 was just one example of utilities on the verge of disruption fighting to keep their cozy monopolies. Utilities in California and other states and countries know that the best way to increase their cash flow is the good-old fashioned way — in the closed-door meetings of state legislatures, public utility commissions, and regulatory bodies.

## The Empire Strikes Back Again: Taxing the Sun

Imagine that your local landline telephone company (let's call it Ma Bell) wanted to charge you a fee simply because you have a cell phone. You receive the following email from Ma Bell:

> Dear [name here]: We have noticed that you have a new cell phone. When you use your new mobile phone, we make less money on our expensive and obsolete infrastructure. We need to compensate for that by charging other "pure-landline" users more for our services. Certainly, we could invest in upgrading our technology and offering services that customers really value, like cell-phone services, but as a monopoly provider, we don't have to. We prefer to charge you a "cell-phone fee" of $50 per month. We do value your money. You'll understand that we are advocating for the millions who do not have cell phones. For instance, our CEO only makes $11.4 million per year, our CNO $9.1 million, and our COO $5.4 million.[185] As such, we have asked the state public utility regulators to promptly approve a "cell-phone" fee. For more information click here.

You would think that the email was a scam and delete it. Even if you thought the email came from the phone utility, you would assume that no government agency would consider, let alone approve, such as scheme.

Except in energy. Substitute "solar" for "cell-phone" and you will see that scenario playing out throughout the United States.

In Arizona, the local utility, Arizona Public Services (APS), asked the Arizona Corporation Commission (ACC), the state public utilities commission, for the right to charge a $50 monthly fee to solar users on the grounds that solar users purchase less power from APS.

Having learned how the cell phone swiftly made the landline phone system obsolete, the power utilities want to have it both ways: they want to milk their obsolete infrastructure through their monopoly power and at the same time use the regulatory system to profit from the transition to a distributed power infrastructure by taxing it.

You'd think the ACC, whose mission it is to make decisions in the public interest,[186] would laugh APS out of the room — but you'd be wrong. On November 19, 2013, the ACC approved a motion allowing APS to charge a fee of at least $4.95 per month to users who have rooftop solar panels. The larger the solar system, the larger the fee APS could charge.[187]

This utility ruling may well leave the doors wide open to similar rulings in other industries. Are you watching YouTube or Netflix? The local cable company requires a $4.95 monthly fee for not using its copper cables. Do you drive an electric vehicle? The oil companies want $4.95 per month for not using their pipelines. Do you cook on an electric stove? The natural gas industry wants $4.95 per month for not using its pipelines to your home.

Watch out for that old telegraph company wanting their piece of your wallet.

# Chapter 4:
# The Electric Vehicle Disruption

*"I do not believe the introduction of motor-cars will ever*

*affect the riding of horses."*

*- Scott-Montague, UK MP, 1903.*

*"The next 20 years of technology change will be*

*the equivalent of the last 100 years."*

*- Ray Kurzweil.*

*"The best way to predict the future is to invent it"*

*- Alan Kay.*

On November 11, 2013, Motor Trend magazine announced the winner of Motor Trend's 2013 Car of the Year award. It was the Tesla Model S.[188]

A company that did not even exist ten years before had built the first electric vehicle to ever win this prize.[189] Tesla fashions itself a Silicon Valley computer company, closer in spirit and thinking to Apple and Google than to its Detroit forebears. Elon Musk, Tesla's CEO, did not deny the possibility of his company being acquired by a "deep pockets" company like Apple.[190]

Prizes are one thing, but when the rubber hits the proverbial road, unit sales are what matter in the auto industry. When market figures for the second quarter of 2013 were released, Tesla had outsold Mercedes Benz, BMW, and Audi in the "large luxury" category, according to Green Car Reports.[191]

Tesla sold just 4,750 cars (an annualized rate of 19,000 units) in an industry that sold 82 million units in 2012.[192] This is a small number of cars in a small car category, so most gasoline car executives probably didn't lose any sleep over the Silicon Valley upstart.

But soon Tesla's stock quadrupled, giving the company a market valuation of $17 billion dollars, more in line with a Silicon Valley high-tech company than a Detroit car company. Tesla had one percent of Ford's revenues but a quarter of its $68 billion market capitalization and a third of General Motor's $51 billion market cap.[193]

General Motors CEO Dan Akerson promptly "ordered a team of GM employees to study Tesla and the ways it could challenge the established business model."[194] Translation: How can Tesla disrupt the $4 trillion global automotive industry?

## Nine Reasons Why the Electric Vehicle

## Is Disruptive

General Motors CEO Dan Akerson had a lot to worry about. Soon after Tesla's stock quadrupled, the Model S achieved the highest safety rating of any car ever tested by the National Highway Traffic Safety Administration (NHTSA).[195] Then Consumer Reports called the Tesla Model S the best car it ever tested.[196]

As I walked through the SF Auto Show on December 2, 2013, I thought of how little the industry had changed in a century. After admiring some of the latest designs like the BMW 950 and the Audi R8, I went to the "museum" section where beauties like the 1928 Hispano-Suiza H6C, the 1931 Cord L29

Convertible Coupe, and my favorite "oldie" of the show, the 1937 Lagonda LG45 Touring Rapide were on exhibit.

The Lagonda LG45 was powered by an overhead valve high-compression 4.5-liter 6-cylinder engine with dual side-draft carburetors mated to a Meadows 4-speed manual gear-box. It was in the museum section of the auto show, but the LG45 looked right at home with the latest Buicks, Volkswagens, Toyotas, and Kias. It was obvious by looking at the 2014 Buick 3.6 liter VVT DI (see Figure 4.1) that the internal combustion engine automobile hasn't really changed much since the Lagonda LG45 engine came out eight decades ago.

On December 10, 2013, the General Motors board of directors announced that it had elected product development chief Mary Barra, a 33-year company veteran, the new company CEO.[197]

Losing their jobs is just one of the reasons why Barra and other conventional car executives should be losing sleep. The internal combustion engine (gasoline and diesel) auto industry is the equivalent of the horse and carriage industry of a century ago. The electric vehicle will disrupt the gasoline car industry (and with it the oil industry) swiftly and permanently.

Figure 4.1—The internal combustion engine. (Photo: Tony Seba)

There are many reasons why the electric vehicle is disruptive. Market disruption will soon make the internal combustion engine (gasoline or diesel) a thing of the past. Even worse from the standpoint of gasoline and diesel cars, the EV is not just a disruptive technology; the whole business model that the auto industry has built over the past century will be obliterated.

The San Francisco Auto Show of 2030 will be quite different from the 2013 version. Even today's impressive BMW 950 and Audi R8 will be the equivalent of beautifully designed horse carriages.

There are nine reasons why the electric vehicle is disruptive.

## 1. The Electric Motor Is Five Times More Energy Efficient

Of the major energy users in the U.S., the transportation sector is the most wasteful. A full 79 percent of petroleum energy used in transportation goes up in smoke (see Figure 4.2). On average, only 21 percent of the gasoline or diesel (both of which are petroleum-based) that is pumped into an internal combustion engine turns into useful energy.

Certainly, as car companies say, mileage may vary depending on driving conditions, the condition of the engine, whether you're driving on city streets or a highway, and so forth. But the numbers speak for themselves and the conclusion is clear: The internal combustion engine is inherently inefficient.

When you consider both city and highway driving for an average car in the U.S., only 17 to 21 percent of the source gasoline energy is used to actually move the wheels, according to the U.S. Department of Energy (see Figure 4.2). A century of knowledge gained from building billions of cars and investing hundreds of billions of dollars in research and development has given us an internal combustion engine vehicle with an energy efficiency of about 21 percent.

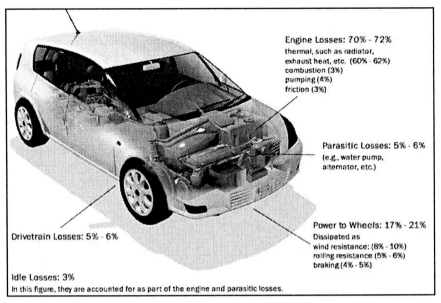

Figure 4.2—*Fuel economy: Where the energy goes. (Source: U.S. Department of Energy)*[198]

Can conventional car manufacturers in Detroit, Munich, and Japan build more energy-efficient engines? Sure. They can make incremental improvements, but the laws of physics stand in their way.

Combustion engines are heat engines and, as such, are subject to the laws of thermodynamics. Specifically, they are subject to the law of maximum thermal efficiency, the upper boundary of how much heat can be turned into useful work. Gasoline engines have an upper limit of about 25 to 30 percent when used to power a car.[199] That is, even at its theoretical best, a gasoline engine will still waste 70 to 75 percent of the energy.

How does this compare with the electric motor? Start by remembering what the electric vehicle does not have: radiators, pistons, an exhaust, a crankshaft, a clutch, pumps, and myriad other thermal engine necessities that waste so much energy (see Figure 4.2).

An electric motor has an energy conversion efficiency of up to 99.99 percent.[200] Tesla's first-generation electric vehicle, the Roadster, had an overall drive efficiency of 88 percent.[201] This figure is four to five times the energy conversion efficiency of an average American gasoline car. The electric motor produces not just a smoother ride, but a far more energy-efficient ride.

Can Detroit design an internal combustion engine that is as energy efficient as the electric motor? In a word: no. The laws of thermodynamics prevent this from happening. Notice what technologies Detroit is using in order to make gasoline cars more energy efficient: batteries and electric motors!

Because the internal combustion engine is a heat engine, you need to increase the temperature of the engine to achieve higher energy conversion. Even coal and nuclear power plants, with their far higher temperatures, waste two-thirds of their potential energy.

To give an idea of how energy-efficient electric vehicles are, the Tesla Model S electric engine is three times more energy-efficient than even large multi-billion dollar nuclear or coal plants!

### 2. The Electric Vehicle Is Ten times Cheaper to Charge

It costs $15,000, or $3,000 per year, to fill up a Jeep Liberty over five years, according to *Consumer Reports*.[202] That's assuming you drive 12,000 miles per year. To drive 12,000 miles in a Tesla Roadster, it costs $313.

Here's the quick calculation: The Roadster goes 4.6 miles per kWh. The average retail price of electricity in the U.S. is 12 ¢/kWh, according to the Department of Energy. Now do the numbers: (12,000 miles * 0.12 $/kWh) / 4.6 miles/kWh = $313.04.

Over five years an Electric Jeep Liberty (if one existed) would cost you $1,565 in electric energy costs instead of $15,000 in gasoline energy costs.

It is about ten times cheaper to charge an electric vehicle than fill up an equivalent gasoline vehicle for two reasons:
- Electric motors are four to five times more energy-efficient than gasoline vehicles.
- Gasoline is two to three times more expensive per energy unit than electricity.

Electricity and gasoline costs vary widely in the United States and around the world. Mileage may vary, as the industry likes to say, but this gives you an approximate idea. In the U.S., an electric vehicle will save about 90 percent in fuel costs per year compared to a gasoline vehicle.

These Jeep Liberty gasoline savings are equivalent to about two years of in-state tuition at Florida State University.[203] So here's a question many families will soon ask themselves: Do we keep burning those dollar bills with expensive gasoline or do we get an electric vehicle and save enough cash to send the kids to college?

### 3. The Electric Vehicle Is Ten Times Cheaper to Maintain

Conventional cars are supposed to have an oil change every three to five thousand miles. There's no need for that with an electric vehicle (EV). But it's not just oil changes that make the EV superior.

Because the EV is powered by an electric motor, it doesn't need any of the parts that have to do with combustion: no spark plugs, starters/alternators, fuel injector, combustion chamber, pistons, piston crown or cylinders, filters, or exhaust. The electric vehicle has no crankshaft or timing belt or catalytic converter. Because an electric vehicle has fewer parts, it is not nearly as needy as the internal combustion engine car. Fewer parts need to be loaded on the chassis (see Figure 4.3).

Data is limited, but it's safe to say that electric vehicles need 90-percent fewer repairs and maintenance work than gasoline/diesel engines do. Consequently, an EV driver spends 90-percent less money on repairs during the lifetime of the car.

*Figure 4.3—The Tesla Roadster chassis. (Photo © Tony Seba)*

### 4. The Electric Vehicle Will Disrupt the Gasoline Car Aftermarket

In 2010 there were 257,576 light vehicle repair shops in the United States.[204] These shops broke down as follows: 3,978 department store locations performing services; 16,800 vehicle dealer repair locations; 77,674 general repair shops. These shops perform all kinds of maintenance and repairs on the nearly 250 million light vehicles on American roads.

Total car manufacturer aftermarket revenues in 2010 were $83 billion; revenues are expected to grow to $98 billion in 2017, according to Frost & Sullivan.[205] Aftermarket costs include repair parts for things like carburetors, spark plugs, starters/alternators, filters, and exhaust components.

But guess what? The electric vehicle doesn't have any of these parts. Remember those engine oil changes every three to five thousand miles? There's no need for that with an EV.

Ultimately it's not just technology that destroys industries; it's the fact that the disrupting companies have business models with which the incumbents cannot compete.

Remember Kodak? They didn't just lose sales of film to digital camera owners. They lost the whole aftermarket for developing film: equipment, paper, and chemicals.

Aftermarket income is an essential component of the conventional auto industry business model. Car manufacturers make an additional $25.8 billion selling tools and equipment to the aftermarket value chain. Aftermarket sales include products such as engine power tools, drain snakes, cutters, and fluid management equipment.[206]

Car manufacturers will not only lose unit sales to electric vehicles; they'll lose the bulk of their aftermarket revenues. They will no longer sell tools and equipment to repair shops or the aftermarket parts needed for repairs.

The internal combustion engine car aftermarket will collapse. The disruption of the conventional auto industry will be complete.

## 5. Wireless Charging

In New York in 1891, Nikola Tesla publicly demonstrated the world's first wireless transmission of energy by electrostatic induction.[207] Tesla went on to invent many of the key technologies and concepts that underlie our electric power infrastructure.

Fast-forward 120 years and you'll find electric buses in Italy being recharged at bus stops while passengers get on and off (see Figure 4.4).[208] The electric vehicle industry has adapted induction power transfer (IPT) so cars can be charged wirelessly without a "typical" charging infrastructure.

*Figure 4.4—Wireless charging of an electric bus in Italy. (Photo Source: Conductix-Wampfler)*[209]

Being able to recharge this way is important for city buses that follow the same route and stop at the same places every day. Instead of one huge meal for breakfast, these buses munch on electricity hundreds of times a day. Wireless charging essentially untethers electric vehicles. Because they are replenished many times a day, they can use a smaller battery. A bus with a 240 kWh battery operating 18 hours per day can go the same distance as one with 120 kWh that is recharged using inductive charging, according to Conductix-Wampfler, the company that built the inductive chargers used in Italy. Using a smaller battery cuts the cost of running the bus by $100,000 or so (at current battery prices).

The same wireless technology that is used to charge a cell phones can be used to charge an EV on the go. General Motors offers Powermat with some of its 2014 vehicles. Powermat uses induction transfer technology to charge smartphones wirelessly. Maybe someday GM's will use this same technology to charge electric cars.[210]

Think of the potentially new business models that induction power technology could open up. Instead of Fedex and UPS vans double-parking and blocking downtown streets, the vans could park at induction charging stations while the drivers deliver packages. Traffic gets better, someone makes money by selling electricity to Fedex and UPS vans, and Fedex and UPS save money by buying electric vans with smaller batteries.

## 6. The EV Has a Modular Design Architecture

The standard architecture for cars has consisted of a single engine that powers two wheels or four wheels via a combination of a transmission, a differential, and drive shafts. The first generation of electric vehicles used this single-motor architecture. However, as electric automakers have developed confidence in their engineering prowess, they have started to take advantage of something that the electric motor has but the fossil-fuel engine cannot compete with: modularity.

The Tesla Model X and the Audi eTron Allroad will have two electric motors, a rear drive unit to power the rear wheels and a front drive unit to power the front wheels (see Figure 4.5).

Like an Intel processor with multicore architecture, this modularity allows for increased power, more design flexibility, and more traction control. It also makes the EV more secure. Electric motors last far longer than internal combustion engines. However, should one of the two motors fail, an electric car does not have to stop running. The second motor can still power the car. And you don't have to stop at two motors. You can have four electric motors, one driving each wheel.

*Figure 4.5—Tesla X dual electric motor (front drive unit and rear drive unit). (Photo Source: Tesla Motors)*[211]

An MIT team led by late professor William J. Mitchell created a novel technology called robot wheels.[212] Each robot wheel has an integrated electric drive motor, steering motor, suspension, and braking system. A car designed with four of these self-contained wheels has a level of flexibility that no commercial gasoline car could ever have: an unencumbered chassis without transmission, gearboxes, and differentials.

Just plug the wheels at the four ends of the chassis, connect them to a battery (or two batteries or four batteries), and control them electronically with by-wire systems.

Having four independent electric motors allows for an extra degree of power, design flexibility, and traction control, said Masato Inoue, chief product designer of the world's best-selling electric vehicle, the Nissan Leaf. Speaking to my class at Stanford University, Inoue showed how the Pivo 3, a Nissan concept car, could park sideways by turning its wheels at 90 degrees. (For a video of the Pivo 3, go to http://youtu.be/J2Ruxm_JdCU.)

The design possibilities are endless. The internal combustion engine could never compete with this.


## 7. Big Data and Fast Product Development

By June of 2013, Tesla Roadsters had driven a combined 30 million miles; the 11,000 Model S cars on the road had driven an additional 30 million miles. Tesla cars had driven a total of 60 million miles (100 million Km).[213]

So what? GM cars have driven trillions of miles. Does Detroit really have a reason to worry?

The difference between the value of Tesla's electric miles and GM's gasoline miles has to do with data. Think of an electric vehicle as a mobile computer. EVs generate vast amounts of data. A car company sifting through this data can learn and adapt far more quickly than a company without user data. A car company that collects data about its cars can understand customer usage patterns and technology stresses and failures. It can quickly fix mistakes, download new software to the cars, and develop new products and services.

The electric vehicle product development process is shortened; it resembles the computer industry's ultra-fast development cycles. Tesla and the other EV companies will be on a product development cycle at the exponential speeds of Moore's Law; Detroit will be on conventional linear speeds.

Within Silicon Valley, you now have to compete at speeds faster than Moore's Law. Take, for example, Apple and its disruptive iPhone. When Apple CEO Tim Cook announced the new iPhone 5s in 2013, he said that the iPhone's central processing unit (CPU) performance had improved by a factor of 40 since its introduction in 2007.[214] The iPhone CPU had improved at an annualized rate of 85 percent! Moore's Law, at a mere 41 percent, pales by comparison. No wonder companies like Blackberry and Nokia have bitten the dust. Even Moore's Law is not fast enough in the ultra-competitive smartphone market.

Apple's iPhone nearly doubles its performance every year while its competitors take two years to double their performance.

This is what Detroit is up against. In the time it takes Detroit to develop a new gasoline car, Tesla could develop two generations of electric vehicles. No industry can survive when its competitors have this kind of development advantage.


## 8. Solar and Electric Vehicles Are Four-Hundred Times More Land Efficient

After the British Petroleum Gulf oil spill of 2010 I wondered about the combination of EVs and solar. At some point every car will be electric; most power will come from solar and wind. What if every car in America were electric and they were all powered by solar?

How much surface area would be needed to generate enough energy to power every car-mile in America with solar power? Americans drive around three trillion miles (4.8 million Km) per year, according to the U.S. Department of Transportation.[215] The Nissan Leaf, the best-selling electric vehicle today, runs 3.45 miles per kWh of battery storage. Imagine building one huge solar power plant in the Nevada, Arizona, or California desert. How large would it have to be to generate enough solar power for all the cars in America? The answer is: 875 square miles.  That is, one square with four sides, each 29 miles long.

Of course, no one will build a solar plant that large. For one thing, it's better to generate power close to the cars themselves — on residential and commercial rooftops, in parking lots, on landfills, and so on. One thousand square miles gives an idea of the magnitude of the surface area that would be needed.

Consider that Wal-Mart is expected to cover 218 square surface miles by 2015.[216] Wal-Mart alone could power nearly one fourth of all electric car-miles in the United States.  All Wal-Mart would need to do is cover its rooftops with solar panels and its parking lots with solar canopies.

The oil and gas industries lease 143,000 square miles of land from the U.S. government to meet about a third of America's transportation needs (see Chapter 7). Multiply that times three and you get 429,000 square miles, the amount of land the oil and gas industries would need to generate all the oil to power every single gasoline (and diesel) vehicle-mile in America.

The combination of solar and the electric vehicle would need 875 square miles, while the combination of oil and the internal combustion engine would need 429,000 to power every single car-mile in America.

The combination of solar and EVs is 490 times more land-efficient than the the conventional car and oil industry.

A technology convergence (solar plus EV) that uses 400 times less valuable resources than existing conventional use (oil plus internal combustion engine cars) is bound to be disruptive. This is especially the case if those resources are as valuable as land and water. It's just a matter of time before the solar-electric vehicle disruption happens.


## 9. Electric Vehicles Can Contribute to Grid Storage and other Services

Imagine being paid to own a car. A recent report by the California Public Utilities Commission concluded that EV owners could be paid up to $100 per month for providing electricity to the electric grid infrastructure.[217]

A gasoline car's integration into the energy infrastructure consists of going to the gas station, pumping gas, and paying. Energy flows one way (into the car) and money flows the other way (to the oil companies.) Electric vehicles can be tightly integrated into the electric grid and provide two-way energy flow as well as cash flow (see Figure 4.6).

Electric vehicles don't have to just sit there and plug into the grid. They can play a dynamic part in the energy infrastructure by providing useful and economically beneficial services to the grid.

An example of how EVs can benefit an electric grid is found in Internet architecture. Companies like Skype became wildly successful by using peer-to-peer (P2P) communications technologies. Each computer with a Skype connection became an active participant, contributing storage, bandwidth, and processing power to the entire network.[218] This decentralized architecture lowered the cost of making phone calls for all participants. A large, centralized infrastructure was no longer needed to handle the phone, text, video, and file traffic.

Electric vehicles can play a role on the grid similar to the role that computer nodes play in a P2P architecture. Electric Vehicles can store energy when there's excess energy production. For instance, the wind generates power mainly at night and in cold weather. In most industrialized economies, more power is consumed during the day; the peak demand for power occurs on the hottest days.

*Figure 4.6—Electric vehicle charging stations at SAP in Palo Alto, California. (Photo Tony Seba)*

Electric Vehicles can release energy during these peak demand times. During hot summer days, for example, EVs can act as on-demand energy providers. Because utilities charge maximum peak prices during these times, EV owners could get paid to provide the energy they store in their EV batteries to the grid. In a competitive energy market every EV owner could participate in energy auctions and sell energy to the highest bidder.

Being paid, say, $100 per month for performing grid services would change the economics of car ownership.[219] A car would go from an expense-only item to an income-producing investment.

## How Long Before the Disruption Occurs?

On April 30, 2013, General Motors CEO Dan Akerson said that the 2014 Chevy Volt, a plug-in hybrid electric vehicle, would cost $7,000 to $10,000 less to build than the previous model.[220] This 20-percent price cut occurred at a time when GM had sold about 26,500 Volts.

Electric vehicle costs are dropping fast, as is the cost of solar panels. Solar panels have a 22-percent learning curve. That is, PV costs have historically dropped by about 22 percent with every doubling of industry capacity.

Learning curves are an integral part of manufacturing industries. Basically, the more cars, panels, or computers a company makes, the more efficient they become at making them and the cheaper the final production unit becomes.

In the automobile industry, the assumption is that batteries must come down in cost before the electric vehicle can match the purchase price of a gasoline car and compete in the marketplace. Nothing could be further from the truth.

Cell phones never really matched the cost of landline phones yet they managed to disrupt the landline telephony industry. Last I checked, a landline telephone costs $10 or $20, while the Apple iPhone costs $600. The cell phone is a re-invention of the old conventional phone, not just a substitution. The cell phone offers value (mobility, for example) that the landline phone can't match; the smartphone has turned into the center of our social lives.

The innovation that enabled cell phones to disrupt the landline phone industry in the U.S. was a business model innovation: the customer committed to a two-year service contract and, in exchange, the carrier financed the phone over that period of time with no down-payment required.

The innovation that enabled the internal combustion engine car to disrupt horse carriages was also a business model innovation: car loans. By the time cars dropped in price to match the horse carriage, the disruption was pretty much over.

Similarly, by the time the electric vehicle matches the capital cost of the gasoline car (sometime before 2030), the disruption of the gasoline car industry will be pretty much over.

When a consumer asks "How much car can I afford?" she is asking what the monthly payment is that fits her budget. Electric vehicles just need to match or be close to the monthly payment of a gasoline car.

The EV industry may be closer to achieving this monthly payment than you imagine. The Nissan Leaf can already be leased for $179 per month (see Figure 4.7).

*Figure 4.7—"100% Electric Vehicle – Lease me for $179 per month plus tax." (Photo Tony Seba)*

Gasoline car aficionados will shout "range anxiety" faster than you can say "horse carriage." The biggest knock on electric cars is you can't drive them very far without a recharge. However, the average American commutes just 29 miles (46.7 km) to and from work. The Nissan Leaf's 75-mile (120 km) range on a single charge is good for a round-trip to and from work plus an extra 17 miles (27 km). Is "range anxiety" real or an ICE industry invention? Either way, I agree with Dan Galves of Deutsche Bank who says that a range of 200 miles (320 km) might be the minimum requirement of a market-leading electric vehicle today. This would leave 142 miles (229.5 km) of spare battery range on top of the average 58-mile (93 km) round-trip commute.

The electric vehicle industry has been quite innovative on the technology front, but it hasn't exploited what may be its most powerful tool: business model innovation.

## Disruptive Business Model Innovations

There were dozens of Internet companies when Google started. Its technology was brilliant, but it was a business model innovation (AdWords) that enabled Google to become the undisputed search-engine leader. Technological innovations create the environment for disruption to happen, but, to make disruption happen, business model innovation is sometimes more important than technological innovation.

Electric vehicles are connected, networked, mobile devices. They can enable new business models that gasoline car vendors simply can't replicate.

## Free Electric Charging

On January 26, 2014, John Gleeney and his 26-year old daughter Jill completed a drive across the United States in John's Telsa Model S.[221] How much did the Gleeneys spend fueling their Tesla electric vehicle? Zero.

Tesla is building the "Tesla Supercharger Network" in North America and Europe so that Model S owners can drive long distances — and drive for free (see Figure 4.8). An important reason for building this network is to counteract the oft-repeated media assertion that electric vehicles have a short range and are therefore not ready for the big leagues. Gleeney's Tesla Model S has an 85 kWh battery with a range of 265 miles (426 km) per charge.[222] The Gleeneys recharged their Model S at 28 charging stations on their cross-country trip. It takes about forty minutes to charge 80 percent of the battery.

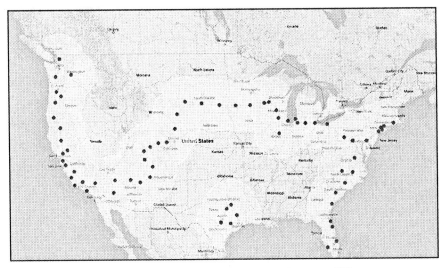

*Figure 4.8—Tesla Supercharger Network. (Source: Tesla Motors)*[223]

But the most important thing about the Tesla Supercharger Network is that it allows Tesla to test a new, potentially disruptive business model. Imagine an EV company that offered free "fuel" for five years or 60,000 miles. Detroit can't afford to do that.

I mentioned earlier in this chapter that it costs $15,000 to fill up a Jeep Liberty over five years, or $3,000 per year, according to *Consumer Reports*.[224] That's assuming you drive 12,000 miles per year. A gasoline car manufacturer couldn't possibly afford to offer free fuel ($15,000 over five years) as part of the vehicle purchase price.

It would be fairly easy for an EV vendor to offer free electricity for five years as a purchase incentive. In the U.S. this would cost the car vendor about $1,500.

Conventional car manufacturers spend $3,000 per car on consumer and dealer incentives, plus $1,100 on media and marketing activities, according to Accenture.[225] That's a total of $4,100 per car. An EV manufacturer offering a $1,500 incentive would be spending 62-percent less than the amount Detroit spends to get a customer today.

Which is worse, "range anxiety" or financial insecurity brought about by high gasoline prices? What percentage of the market would be willing to drive an EV that included free charging?

Including free charging with every EV would be a disruptive business model. Moreover, this business model would obliterate the gasoline car industry; they could do nothing to compete with it. Absolutely nothing. Once this business model becomes the industry standard, the oil age will basically be over.

## Free Maintenance

Electric vehicle (EV) companies could make the demise of the internal combustion engine industry happen even faster by offering free maintenance.

EV maintenance costs are an order of magnitude lower than internal combustion engine costs. An electric motor can last for decades, but a heat engine breaks down time and again. The EV doesn't have hundreds of parts — carburetors, spark plugs, starters/alternators, filters, and exhaust components — that need constant care.

Imagine an EV company that offered free maintenance for five years or 60,000 miles. This would be another disruptive move that ICE companies couldn't match.

The aftermarket business is actually a big cash cow for car manufacturers today. Turning a major revenue line item like the aftermarket business into an expense line item would probably send most car manufacturers into bankruptcy.

Earlier in this chapter I explained nine reasons why the electric vehicle is disruptive. As we speak, two guys or gals in a Silicon Valley garage may be at work on new disruptive business models that Detroit won't be able to compete with.

## My 2010 Prediction for the End of Gasoline Cars by 2030

In the summer of 2010, I gave a keynote speech in Dickinson, North Dakota, where I predicted that gasoline cars would be obsolete by 2030. (The video is here: http://youtu.be/MAFoqo3Jbro). North Dakota was going through a full-fledged fracking "revolution" and was well on its way to becoming the second largest oil producer in the nation. Meantime, Tesla Motors had barely delivered one thousand of its first-generation electric vehicles, the Roadster. My prediction sounded positively crazy, but it may well have been conservative.

In 2010, the rule of thumb was that EV-quality Li-on batteries cost about $1,000 /kWh. The Tesla Roadster had a 53 kWh battery, which cost about $53,000, about half the total cost of the car.

To make my prediction I made what I thought was a reasonable assumption. I assumed that Li-on batteries would fall at a 12-percent yearly rate (see Figure 4.9). At that rate, the cost of Li-on batteries would reach $100/kWh by 2028. When EV storage prices reach that threshold, I told the audience in North Dakota, it's "game over" for gasoline cars (and for oil). My prediction was greeted by a long silence.

While everyone else promised a new millennium for oil and gas, I told my audience that the reign of oil would be finished in two decades. Either the energy consensus was wrong or my prediction was crazy. On my way back to the airport I overheard the speaker who followed me, a politician, talk about the need for "clean coal." I didn't have to wonder which of us was the crazy one.

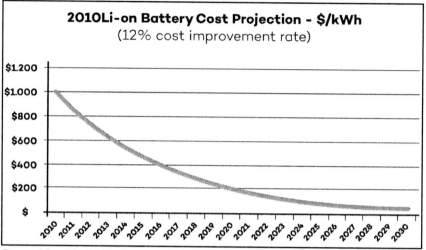

*Figure 4.9— My 2010 electric vehicle Li-ion battery yearly cost projection in $/kWh. (Tony Seba)*

The most expensive part of today's electric vehicle is the battery. The most popular EVs, such as the Tesla, Nissan Leaf and the Chevy Volt, use Lithium-Ion (Li-on) batteries. The Tesla Model S base model comes with a 60 kWh battery. Its estimated mileage is 230 miles (370 Km).[226] That is, it can go approximately 3.83 miles per kWh of energy stored in its battery. The Roadster had a 53 kWh battery and a range of 244 miles; it could go about 4.6 miles per kWh.

For my calculations I'm using an average of four miles range per kWh of battery storage. Basically, an EV with 200-mile range would need a 50 kWh battery. Notice also that not all Li-on batteries are equal. As in many product lines (smart phones, shirts, or cars), there's a range of quality and price points to choose from.

When Li-on reaches $100/kWh, the batteries for an EV with a range of 200+ miles will cost about $5,000. Assuming that the battery is about a third of the price of an EV, you'll be able to purchase an entry-level equivalent of a Tesla Model S for $15,000.

By comparison, the average cost for a new vehicle in the U.S. in 2013 was $31,252.[227] Even the low-end vehicles, such as the Huyndai and Kia brands, sold for an average $22,418.

GM's operating margin was 3.3 percent in 2013, up from -19.9 percent in 2012, according to Morningstar.[228] Based on those figures, GM does not have much pricing room without getting into negative, financially unsustainable cash-flow territory. Premium brand manufacturer BMW had operating margins in the 8.7

to 11.7 percent range from 2010 to 2012.[229] This gives BMW a little more room to maneuver, but not much room. A 10-percent price swing in the market could put both companies solidly in the red.

The gasoline industry will simply not be able to compete with a Tesla Model S-quality vehicle that retails for $15,000 or $20,000. Not Kia, not GM, not Toyota, not BMW. The gasoline car industry will be in trouble when storage batteries reach $100/kWh. At that point, the disruption will be long over.

My initial projections in 2010 pointed to the industry reaching $100/kWh in 2028. Now it looks like it will happen even sooner.

My 2010 prediction was in the "ballpark," but electric battery costs are going down a little faster than I predicted in 2010. As is usually the case in technology markets, the virtuous cycle of innovation, competition, and scale is pushing Li-ion battery costs down a bit more quickly than anticipated. "The cost of manufacturing Li-on batteries has dropped from $1,000-$1,200 to $600 in just three years [from 2009 to 2012]," said the secretary of the U.S. Department of Energy in 2012.[230]

I was not far off. My original prediction said that, by 2014, EV-Li-on batteries would be around $600/kWh. Today, EV batteries are in the neighborhood of $500/kWh.

## My New Prediction for the End of Gasoline Cars by 2030

Tesla's car batteries are composed of thousands of small Li-on cells similar in size to the Li-on cells in laptop computers. It makes sense that the cost curve for EV Li-on batteries should approximate that of laptop computer Li-on batteries.

How much further will the cost of Li-on batteries fall? We can look to a recent historic precedent to guide our thinking in making this prediction: the cost of Li-on batteries to power laptop computers, smartphones, and tablets. According to Deutsche Bank, laptop computer battery costs fell from $2,000 to $250 over a period of about fifteen years. This represents a cost improvement rate of 14 percent per year.[231]

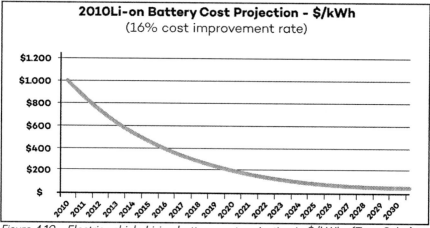

Figure 4.10—Electric vehicle Li-ion battery cost projection in $/kWh. (Tony Seba)

Based on this yearly cost improvement rate of 16 percent (see Figure 4.10), my 2014 cost projection for Li-on batteries is $498 per kWh in 2014. This fits the data better. I love the scientific method. Evidence-based data should drive hypotheses, not the other way around.

It makes sense that Li-on battery costs are falling faster than the historical precedent. Never has investment in electricity storage been so high. At least three multi-trillion dollar industries are now investing billions to come up with better batteries: electronics, automotive, and energy. Apple, Samsung, and Google are as interested in batteries as Tesla, SolarCity, and General Electric.

Tesla recently announced a $5 billion next-generation battery factory (dubbed "GigaFactory") in the U.S. This investment would singlehandedly double the world's Li-on battery manufacturing capacity. The factory will open around 2017 and produce enough batteries for 500,000 cars per year by 2020. Tesla projects it will sell about 35,000 cars in 2014, so the factory would represent a minimum car unit growth of fourteen times in six years.[232] Panasonic, the Japanese electronics giant, was said to be in talks to invest $1 billion in Tesla's "GigaFactory."[233] SolarCity has been using Tesla's batteries as part of its solar panel installations.[234] These batteries allow solar customers to store solar power as well as purchase grid electricity when it's cheap and use the electricity when it's expensive. In fact, the factory itself will get 100 percent of its energy from solar and wind power generated near the factory. Presumably, the factory will make the batteries that will store the solar and wind energy that it will use to manufacture more batteries.

The lines between the auto industry, the energy industry, and the electronics industry are blurring. These lines will soon be nonexistent.

The 16-percent battery cost improvement rate does not even take into account the possibility of major breakthroughs. Tesla management is talking about a "step change in battery technology within five to ten years that would enable 500-1,000 miles of range and full-charging within seconds."[235]

Academic institutions around the world, including my alma maters Stanford University and the Massachusetts Institute of Technology, have made energy one their highest research and development priorities. Their work has already started to bear fruit. MIT Professor Donald Sadoway has focused on the development of liquid metal batteries for grid storage using cheap and widely available materials.[236] Ambri, the first successful spinoff of his lab at MIT, quickly raised $15 million in venture capital from Bill Gates and others; Ambri has already been named one of the "50 Disruptive Companies" of 2013 by *MIT Technology Review*.[237] Stanford Professor Yi Cui's group is using nanotechnology and cheap materials such as silicon and sulfur to build batteries from the ground up using carbon nanotubes, graphene, and other advanced materials. Early results point to order-of-magnitude improvements in cost as well as energy densities for energy storage.[238] An example is a new configuration for Redox Flow batteries that could decrease their production cost to about $45/kWh.

My new electric vehicle Li-ion battery cost projections (see Figure 4.10) fit the recent past (2010–2013) better. Assuming the next dozen years are also governed by the 16-percent yearly improvement in the cost of Li-on batteries, the automotive industry is in for a quick transformation (see Table 4.1).

| Year | 2014 | 2015 | 2016 | 2017 | 2018 | 2019 | 2020 | 2021 | 2022 | 2023 | 2024 | 2025 |
|------|------|------|------|------|------|------|------|------|------|------|------|------|
| Cost ($/kWh) | $500 | $420 | $353 | $296 | $249 | $209 | $176 | $148 | $124 | $104 | $87 | $73 |

*Table 4.1— Electric vehicle Li-ion battery cost projection in $/kWh. (Tony Seba)*

Essentially, Table 4.1 points to EV-batteries reaching $100/kWh in the 2023 timeframe. By 2025, the cost falls to $73/kWh

This looks to be an aggressive price reduction projection. Tesla, however, is already ahead of the curve. While Tesla does not disclose its costs, we can deduce the approximate value of its batteries from published prices. The Model S60 sticker price is $71,070 (not including incentives); the Model S85 sticker price is $81,070.[239] There are four main differences between these two models: the S85 has a larger battery (85 kWh vs. 60 kWh), the S85 has a larger motor (362 hp vs. 302 hp), and the S85 includes a supercharger and a 19-inch Michelin tire upgrade.

The supercharger retails for $2,000 and the Michelin tire upgrade retails for $1,000 for a combined price of $3,000. Assuming a 50-percent margin on these two items (which is quite large), the cost to the company is $1,500. Further assuming that Tesla makes no margin on the battery or the motor upgrade, the incremental 25 kWh of battery storage costs $8,500. This means that the battery would cost a maximum of $340/kWh. This figure may be the upper boundary of Tesla's battery costs. Deutsche Bank published a report in July 2013 that estimated Tesla's batteries cost around $350/kWh.[240]

What do these battery costs mean in terms of the transition to an all-electric vehicle automotive industry? Electric vehicle reports by several influential analysts, including McKinsey, Morgan Stanley and Deutsche Bank, point to the fact that, in the U.S. market where gasoline costs $3.50 or more, electric vehicles become competitive as battery costs reach the $300 to $350/kWh level. In European markets, where gasoline retails for $8 per gallon or more, EVs become competitive when EV battery prices fall below $400/kWh.

Tesla is already in the disruption sweet spot. The highest ranked affordable midsize SUV in America in 2013 was the 2014 Buick Enclave, according to *US News & World Report*.[241] The Enclave retails for $38,698 to $47,742. Tesla's sports-utility vehicle, the Model X, to be launched in 2015, is expected to retail for $35,000 to $40,000, placing it squarely in the American mainstream.[242] The base Model X will have a 60 kWh battery and a range of 265 miles.

The Model X SUV, however, "will have the performance of a Porsche 911 Carrera," according to Tesla CEO Elon Musk. This means that Tesla will offer a $100,000 performance car-quality SUV for the price of a $40,000 "affordable" SUV. The Buick Enclave does not stand a chance. Neither does any other SUV in that price category. Not even Porsche.

If Tesla could manufacture millions of cars per year, the mass migration to electric vehicles would start with the Model X. However, despite their technology prowess, design capabilities and market success, Tesla has put a premium on manufacturing quality cars. Tesla's devotion to quality has limited its ability to quickly scale to producing millions of cars per year.

Most large EV manufacturers have not reached the $350/kWh battery cost level that Tesla has. The market seems to be around $500/kWh. Table 4.1 indicates that the market will probably reach the $350/kWh level in 2016 or 2017. This means that the mass migration to electric vehicles will start in 2016 or 2017. It also means that Tesla has a two-year battery cost advantage over the rest of the market.

## The Mainstream Mass Migration to Electric Vehicles

My projections indicate that Li-on batteries will reach $200/kWh by 2020 (see Table 4.1). My projections aren't far from the current consensus. EV batteries will reach $200/kWh to $250/kWh by 2020, according to Anand Sankaran, executive technology leader for energy storage and high-voltage systems at Ford Motor Company.[243] Consulting company McKinsey pointed to $200/kWh by 2020 as far back as July 2012;[244] Navigant points to costs as low as $180/kWh by 2020.

When Li-on reaches $200/kWh, the batteries for an EV with a range of 200+ miles will cost about $10,000. Pushing battery costs down makes the cost of the battery a smaller portion of the cost of making the car. The original Roadster battery cost about half the cost of the car, but Tesla is pushing down battery prices. Currently the cost of the battery in a Tesla is about a quarter of the cost of the car.[245]

Assume for a moment that the battery is a more conservative one-third of the cost of an EV. At a one-third cost, when Li-on batteries reach $200/kWh, you will be able to purchase the equivalent of a 200-mile range Tesla Model S for about $30,000. In fact, Tesla's next-generation vehicle, the Model E, currently planned for 2017, is expected to cost around $35,000.

Again, the average cost for a new vehicle in the U.S. in 2013 was $31,252.[246] This "average" vehicle (Toyota, Ford, GM, Honda, Nissan) will cost about the same as a Tesla EV "with the performance of a Porsche 911 Carrera"; with a sticker price around $30,000. Gasoline cars will not stand a chance. Did I mention that EVs cost 90-percent less to fuel and maintain?

Even low-end vehicles such as the Huyndai and Kia brands, which sell for an average $22,418, don't stand a chance. Would you buy a Kia for $22,000 or an EV with the performance of a Porsche for $30,000?

Again, Tesla is ahead of the curve. CEO Elon Musk said that his company is already looking at a 30- to 40-percent cost improvement (per kWh) over the Model S battery. Assuming that the Model S battery costs $350/kWh, the next generation Tesla battery will come in at $210 to $245 per kWh. Tesla would achieve this cost basis by 2015 or so — several years ahead of the market consensus target year 2020.

# The Last Gasoline Car

When batteries fall to the $100/kWh level, the batteries for electric vehicles with a range of 200 miles could cost as low as $5000. At this level, EVs in all categories become cheaper to purchase than their "equivalent" gasoline cars. I write the word "equivalent" in quotes because today's auto industry classifies cars using an outdated price/performance paradigm. For example, an electric vehicle like the Tesla Model X would belong in the same performance category as the $100,000 Porsche 911 Carrera, but cost the same as a $40,000 Buick Enclave SUV. Electric vehicles will disrupt this obsolete price/performance paradigm.

The conventional car industry is wrong in the way it classifies the gasoline "equivalent" of electric vehicles because EVs are a new breed altogether. "The electric vehicle is a re-invention of the automobile, not just a substitution," said Masato Inoue, chief product designer for the Nissan Leaf. A Porsche cannot compete with the Model X because it will cost more than twice as much as the Model X but deliver an equivalent performance. The Enclave will cost the same as the Model X but will deliver a fraction of its performance.

The electric vehicle changes the basis of competition in the transportation industry (see Figure 4.11). Of all the reasons why the electric vehicle is disruptive, this is probably the most powerful one.

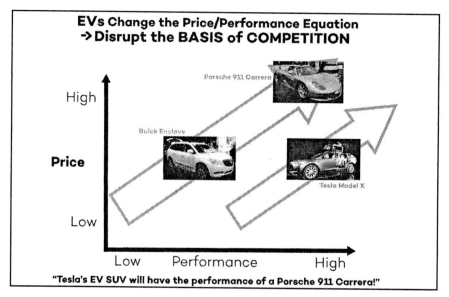

Figure 4.11—Electric vehicles disrupt the basis of competition in the automotive industry. (Source: Tony Seba)

Neither the high-end or low-end gasoline car stands a chance against electric vehicles once EVs are in the same price range. When EV-quality batteries reach $100/kWh, the internal combustion engine industry will be toast. At that point it won't make financial sense to own a gasoline or diesel vehicle no matter the cost of petroleum.

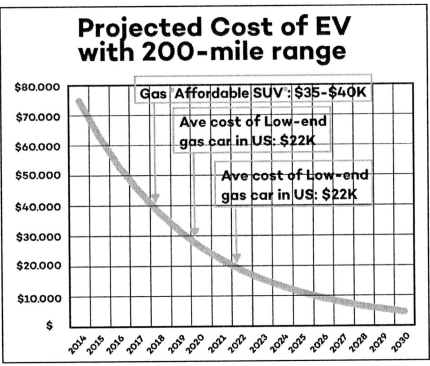

Figure 4.12—Projected cost of an electric vehicle with a 200-mile range. (Source: Tony Seba)

My projections point to the industry reaching the $100/kWh price level by 2024 or 2025 (see Figure 4.12). Starting in 2025, it will make no financial sense to purchase a new gasoline car in any market. At this point most of the high-end and mainstream ends of the market will have transitioned to electric vehicles. Even assuming that my predictions are off by five years or that it takes an extra five years to build out the manufacturing infrastructure and transition to the new EV world order, gasoline cars will be the 21st century equivalent of the horse carriages by 2030.

There may still be millions of older gasoline cars and trucks on the road. Ten- to twenty-year-old cars are still on the road today. We may even see niche markets like Cuba where 50-year old cars are the norm. But essentially no internal combustion engine cars will be produced after 2030. Oil will also be obsolete by then.

Oil will likely be cheaper in 2030 than it is today (see Chapter 8). Still, as the gasoline car industry starts imploding around 2025, the ICE car aftermarket will collapse with it. Electric cars do not need that much maintenance and do not wear out as much. There will be fewer gas stations, repair shops, and used-parts stores to cater to ICE car owners. As ICE car companies go under or move to making electric vehicles, they will stop making parts for their older cars. Just as it's harder to buy camera film or an old Motorola 68000 microprocessor, it will be progressively harder to get parts for older ICE cars. The cost of operating and maintaining a gasoline car will necessarily go up until used ICE car ownership becomes too expensive for all but the most devoted gasoline car owners.

So there's the answer to former General Motors CEO Dan Akerson's question of how Tesla could disrupt the "established business model" of the gasoline car industry. The electric vehicle will disrupt the gasoline car and make it obsolete by 2030, maybe by 2025.

GM thought it killed the electric car in the 1990s, but the electric car may well end up killing GM and all its internal combustion engine siblings as soon as 2025. GM's new CEO Mary Barra, as well as her peers in Detroit, Munich and Toyoda, still have a chance to make it in the new world order of electric vehicles. Given the long product cycles in the gasoline vehicle industry they have to commit to EVs today. Choosing to wait is  choosing to be disrupted.

Whatever is left of the internal combustion engine car industry will be wiped out by another disruptive wave:  the autonomous (self-driving) car.

# Chapter 5:
# The Autonomous (Self-Driving) Car Disruption

*"This 'telephone' has too many shortcomings to be seriously considered as a means of communication."*

*- William Orton, president of Western Union, in 1876.*

*"The significant problems we face today cannot be solved with the same level of thinking that created them."*

*- Albert Einstein.*

*"You have to let it all go: fear, doubt, disbelief. Free your mind."*

-Morpheus, The Matrix.

In 2005 I decided to sell my Porsche Boxster and try a new car-sharing service called Zipcar. For about $6 per hour (or $72 per day), I had access to twenty or so cars parked within a four-block radius at anytime (see Figure 5.1). I just had to a reserve a car on the Zipcar website (the smartphone app came years later), pick it up, and drive it off. There was no need for keys. I just swiped my membership card (ZipCard) and the car door would unlock. The rental price included insurance, gasoline, and free miles (up to 120 miles at a time.)

Back then, the so-called sharing economy was in its infancy. Less than a decade later, the way we think about asset ownership has changed dramatically. Car sharing companies like Zipcar have made it possible for thousands who don't own cars to reap nearly all the benefits of car ownership for a fraction of the cost of owning a car.

Zipcar claimed more than 760,000 members and $270 million in revenues in 2012.[247] According to the company, each of its cars replaced fifteen cars on the road.[248] Under this 15-to-1 share-to-own ratio, Zipcar's 10,000 cars canceled 150,000 potential car sales. To put it another way, Zipcar's car-sharing model may have prevented car manufacturers from selling 150,000 cars. I'm not sure how many big auto executives have lost sleep over the share-to-own ratio of Zipcars to conventional cars, but they soon will lose sleep.

## Cars in the New Sharing Economy

Zipcar was just the beginning of a series of waves that will disrupt the auto industry, the public and private transportation industries, and the logistics world.

*Figure 5.1—A Zipcar car sharing "pod" at San Francisco's Civic Center Plaza. (Photo: Tony Seba.)*

The sharing economy has extended to homes, the average American's most valuable financial asset. Traditionally, our home has also been our castle. Now, with web services such as Airbnb.com, thousands of homeowners make extra cash by renting their homes to total strangers from around the world

An estimated 300,000 people have rented their homes to nine million strangers through San Francisco-based Airbnb.com.[249] Started in 2008, Airbnb has 500,000 listings in 34,000 cities in 192 countries around the world.[250] In less than five years Airbnb went from being an idea in its founders' heads to arguably the largest hotelier in the world.

Even more people have rented their homes using services like CouchSurfing.com. Started in San Francisco in 1999, CouchSurfing.com has allowed six million people to stay in 100,000 cities around the world for free.[251]

There's a difference between sharing services like Zipcar and Airbnb.com. Zipcar is like a hotel franchise for cars. Hotel companies own rooms and rent them on a time-availability basis. Zipcar also owns its cars.

After houses, the automobile is most Americans' largest asset, yet Americans only use their cars about two hours per day. That figure represents less than 10 percent of capacity utilization for what is a pricy asset. We pay hundreds of dollars every month for car loans, insurance, parking, repairs, gasoline, and maintenance — all for an asset that is idle and unused 90 percent of the time.

Is there a way to make money from cars during this "car downtime"? Since we only use cars a couple of hours per day, there is plenty of spare capacity for people to rent cars. After all, if homeowners are happy to share their most valuable financial asset, shouldn't car owners share their second most valuable asset too?

A number of new car-sharing services with a number of slightly different business models have sprung up in San Francisco. Lyft aims squarely at the taxi market. It allows people to use their cars in their spare time as taxi cabs. Everything, from requesting a car to making payments to leaving ratings of drivers, is done with a smartphone app.

Uber started by connecting limousine drivers to potential customers on-demand. The company added an eBay-like auction-pricing model to match limo demand to limo supply. The result is that a peak-demand ride can be much more expensive than a standard taxi ride. I rode an Uber car to a New Year's Eve party in San Francisco's North Beach and paid $50 for a ride that would have cost $15 or $20 in a normal taxi. The driver told me that after midnight rides were expected to cost more than $100.

Lyft and Uber are market-making intermediaries in the personal transportation market. They help make this market more efficient by connecting sellers with spare capacity to buyers who would otherwise not be able to use this spare capacity. These companies have already transformed the taxi market in San Francisco and are expanding globally at a breakneck speed.

Uber was started in early 2009. Less than four years later, it was fielding one million car-ride requests per week and completing about 80 percent of them. The company's 2013 yearly revenues were estimated at $213 million.[252] Uber recently raised more than $341 million of venture capital in a round that valued Uber at more than $3.5 billion.[253] The largest investor in that round was Google, which put $250 million into the four-year-old company.

Taxis, however, are just a small fraction of cars on the road today. More than one billion cars roam the planet, according to Wards Auto.[254] The number of vehicles is expected to grow to 2.5 billion, according to the International Transportation Forum.[255]

Most of the world's billion vehicles are parked in a driveway or parking lot 90 percent of the time. There is an immense market opportunity to the company that can successfully mobilize even a small portion of this idle mobility capacity.

Another peer-to-peer service called GetAround.com aims to make better use of the world's cars. Instead of leaving your car idle for hours every day or days at a time, GetAround asks you to rent it to a neighbor or someone who lives nearby. This service is closer to the Zipcar model in that the consumer can rent a car by the hour or by the day. Unlike Zipcar, however, GetAround does not have to buy and maintain a fleet of cars. The company just connects buyers and sellers and pockets a percentage of the transaction.

Conceptually, the issue with this business model is that your neighbor's car may not be there when you need it. She probably takes it to work in the morning and brings it back in the evening. Your neighbor's car is idle at work 25 miles away when you need it to run to the supermarket. Business models like GetAround.com's work in areas with a high density of buyers and sellers like San Francisco and New York. Can the GetAround.com business model work in a suburb, twenty-five miles away from the action?

# Autonomous Cars:
# The Ultimate Disruption Machine

The self-driving car will disrupt the auto industry, the transportation industry (public and private), and the logistics industry. It will also help disrupt the oil industry. The autonomous car will radically redraw the maps of our cities in a way that hasn't been done since the demise of the horse and carriage.

A self-driving car would pick you up, drive you to your destination, drop you off, and pick up the next customer. When you need to go to your next destination, you would put in an order on your smartphone app (or just ask Siri). Another autonomous vehicle would pick you up and drive you there.

Who owns the autonomous car, a car-sharing company like Zipcar or an individual who rents it out while she's at work, is irrelevant. You know there will be a car to drive you anywhere at anytime. You don't even need to live in a high-density area.

The autonomous vehicle "mobility on demand" business model would expand the transportation market. Think of the millions of people with disabilities, children, and the elderly who can't drive cars. They would have cars to drive them to school, the park, the doctor, or to the homes of family or friends.

Parents who are pressed for time will not need to drive their children to school every morning or their parents to the doctor. The blind could go to a restaurant across town. All this could be done without a licensed driver or a private car.

## Gasoline Cars: The Ultimate Waste Machine

During rush hour, the I-880 Highway in the Bay Area has some of the worst traffic in the United States. When I have to go anywhere near the I-880 I schedule my meetings for the late mornings or early afternoons so I can zip in and out without hitting rush hour traffic. But life is seldom linear even if you just want to go from point A to point B.

I was recently invited to an event at the SFUN solar accelerator where I'm an advisor. The event was at Jack London Square in Oakland at 5 p.m. When I'm home in San Francisco, I take the BART (Bay Area Rapid Transit) underground train or the ferry to Oakland. Alas, I had an early afternoon meeting in San Jose that ran late. When I finally hit the I-880 it was 4 p.m., approaching rush hour.

I had exactly one hour to cover the 38 miles to Oakland, which is normally plenty of time, but the 880 was already like a parking lot. Twenty minutes into my trip I had advanced only five miles.

At that point, I wished I had a self-driving car (see Figure 5,2)

*Figure 5.2—Google's converted Lexus self-driving car. (Source: Wikipedia)*[256]

The world was turning but my car was idle. I needed time to work on this book and my upcoming disruption course at Stanford. I needed to review a design for a wind power plant assessment project. I was an advisor on a potential investment deal in an electric bus company and needed to call my business partner in London. It was already past midnight in the U.K., but maybe I could send my partner an email and he was still awake to receive it. I hadn't called my friend Sara in a while. Or Margie. And, oh yes, I needed time to relax, too.

But I was stuck in traffic. It was clear that I wouldn't make it to the party in Oakland. I turned around and went back to San Jose.

The gasoline car is the ultimate waste machine. The car has brought us waste in at least six different dimensions:
    1. Waste of lives
    2. Waste of time

3. Waste of space
4. Waste of energy
5. Waste of money

The autonomous car is a game-changing product in terms of minimizing every single waste item on this list.

## Waste of Lives

The number of deaths caused by car accidents is a human tragedy of unspeakable proportions.  In the U.S. alone, six million car crashes caused 32,788 deaths in 2010. It is estimated that 93 percent of those deaths were caused by human error. In 2009, 2.3 million adult drivers and passengers ended up in a hospital emergency room.[257]

To put highway deaths in context, 58,220 Americans died in the Vietnam War (1956–1975).[258] During those same years, 757,538 people (thirteen times as many) died in motor vehicle accidents in the United States.[259]

Globally 1.24 million people died of traffic-related accidents in 2010, according to the World Health Organization.[260] Almost half of those deaths were pedestrians, cyclists, or bikers.[261] Additionally, 20 to 50 million people suffer non-fatal injuries because of traffic accidents annually.

Worldwide, death by traffic accident is the leading cause of death for people between the ages of 15 and 29; it is the second leading cause of death for children between the ages of five and 14. More children age five to 14 die in auto accidents than die from malaria, tuberculosis, or measles.[262]

Clearly, humans are not great drivers. We are easily distracted. We drink and eat while driving. We text and speak on the phone. We reach for radio dials and the glove box. We put on make-up, talk to our fellow passengers, try to reason with kids in the backseat, and daydream — sometimes all at the same time. We're also at the mercy of our physical limits. How well we can see, our reaction times, and even our sleep patterns determine how well we can drive a car.

Mathematical models based on Center for Disease Control data imply that drowsy drivers might be involved in 15 to 33 percent of all fatal crashes in the United States, according to Harvard University Professor Sendil Mullainathan.[263]

Autonomous cars can be superior drivers in many ways. They have a 360-degree "visual range." Computers don't get distracted. They can "see" at night and are not at the mercy of sleep patterns.

Their attention doesn't waver because they're texting or talking on the phone. They don't drink, travel at inadequate speeds, or daydream.

Autonomous driving technology is not perfect yet, but autonomous cars may already drive better than most humans. "Autonomous cars are 6,500 times better at detecting danger than humans," said Masato Inoue, chief product designer of the Nissan Leaf. The Google self-driving car has gone 500,000 miles without causing an accident. It has, however, been rear-ended — by a human driver.

Autonomous cars are getting exponentially better because their technological components — vision, sensing, processing, and machine learning — are getting exponentially faster, cheaper, and better. The "artificial intelligence" software that drives autonomous cars is also improving. The amount of data a self-driving car can access is increasing exponentially; the computing platform on which it runs is also improving exponentially.

When a human being learns a lesson, he may or may not share it with others. And even if he shares it, whether others can take heed of the lesson is questionable. A lesson doesn't really sink in unless you actually experience it. People tend to make the same mistakes again and again. This is one reason why there are so many traffic accidents.

By contrast, the Google car gathers more than 1GB of data per second![264] To give you an idea of how much data this is, the iPhone 5 has 16GB to 32GB of data storage capacity. The Google car would fill the data storage capacity of Apple's latest smartphone in 32 seconds or less. Like every other computing platform, data generation is growing exponentially. It won't be long before an autonomous vehicle will be able to fill the data storage capacity of a smartphone every second. What's more, the Google car will "learn" from all this data.

Autonomous cars will also learn from the data collected by all other autonomous cars. An autonomous car that makes a mistake, say, in New Zealand, learns a lesson from the mistake. But what happens in New Zealand doesn't stay in New Zealand. The lesson is codified and transmitted to a "deep learning" database in Mountain View or Munich, where it's catalogued, compared to similar mistakes by other cars, re-codified, and then quickly transmitted to millions of autonomous cars around the world. Within days (or even hours) of the accident in New Zealand, every autonomous car in the world will have learned how to avoid making a similar mistake; every car learns how to be a better driver.

Any accident anywhere will make every autonomous car everywhere a better driver. In information economics this is called network effects. The value of the network increases exponentially as the network acquires more information. Autonomous cars' ability to learn will improve exponentially, which will soon make them smarter, better, faster, and safer than the best human drivers.

That autonomous cars will be better, smarter, and faster drivers than you and me is a given. That they will best the top professional car drivers in the world may seem surprising, but they will leave Danica Patrick, Jimmie Johnson, Dale Earnhardt Jr., and the top NASCAR drivers in the dust. Not only will autonomous beat the top NASCAR car drivers, they will do so while saving fuel and saving lives.

Autonomous cars will soon save more than one million lives a year. This alone is potentially revolutionary.

## Waste of Space

If an archeologist from outer space came to study earth, she would rightly conclude that cars are the dominant life form on the planet. More urban space is dedicated to the car than the human being.

In North American cities, roads and parking lots respectively account for 30 and 60 percent of the total surface.[265] This does not include driveways and garages. Highways are also a massive waste of space. Automobile driving occupies ten to a hundred times more road space than other forms of transportation. For instance, it takes 200 m$^2$ (2,152 square feet) of road space per car passenger versus 30 m$^2$ (323 sqft) for arterial driving, 2 m$^2$ (21.5 sqft) per public transportation passenger, and 3 m$^2$ (32.3 sqft) for walking.[266]

It's easy to think there aren't enough highways when you're going down the road at 60 miles per hour (100 Km/h) and traffic is moving smoothly, but vehicles only use 5 percent of the road surface, according to UC Berkeley Professor Steven Shladover.[267] This means that, at best, 95 percent of the highway surface is not being used at any given time. This space is wasted because, at highway speeds, cars need 120 feet (40 m) to 150 feet (50 m) of space in front for safety purposes. They also need lanes that are twice their width.

Research has shown that intelligent vehicles can dramatically decrease the space needed to perform the same basic functions that humans perform now. For instance, automated vehicles require 25 percent less space for merging and lane changing.[268]

Cars equipped with adaptive cruise control (ACC) can improve highway capacity by about 40 percent. Using both ACC and inter-vehicle communications can boost highway capacity by an astonishing 273 percent, according to research

done at Columbia University.[269]
In other words, autonomous vehicles could end congestion on highways by increasing highway capacity by 3.7 times. After the self-driving car disruption, we will have to decide what to do with all that unused highway space.

## Waste of Time

Traffic congestion cost Americans $121 billion in 2012; this figure is expected to grow to $199 billion in 2020, according to TTI's Urban Mobility Report.[270] Congestion costs Americans 4.8 billion hours of time, 1.9 billion gallons of wasted fuel, and $101 billion in combined delay and fuel costs each year.[271]

While I have mostly looked into cars, anyone who has seen a FedEx or UPS truck double-park knows that trucks also waste time, space, and energy. In 2004, before the explosion in truck deliveries caused by the growth of online commerce, delivery trucks caused an estimated 1 million hours of vehicle delays.[272] A study found that trucks double-parked the equivalent of seven hours per day, turning road space into their own parking places and worsening already-congested cities during peak daytime hours.[273]

Time wasted on the road is also stressful. MIT Professor Carlo Ratti, who developed the road frustration index to quantify the impact of traffic on mental health, concluded that the stress of city driving is as high as that of skydiving.[274]

By decreasing congestion, self-driving cars will also dramatically decrease commuting time. Furthermore, because self-driving cars don't need us to direct them, time we waste today in our cars will be turned into productive time. Some may choose to surf the web (or whatever has disrupted the web by 2030) or sleep in their cars. Either way, time not driving is time added to our lives. Self-driving cars will also save us all the time we waste parking and looking for parking places. They will drop us off and go on their merry way to self-park or pick up the next ride.

## Waste of Energy

In congested urban areas, 40 percent of all gasoline usage is wasted looking for a place to park the car, according to the MIT Media Lab.[275] Congestion costs Americans 1.9 billion gallons of wasted fuel and $101 billion in combined delay and fuel costs. That's $713 per year for each commuter.[276] The first energy-saving feature of the autonomous vehicle is very mundane: self-parking.

When they do need to park, autonomous vehicles (AVs) are more precise at parking; they can squeeze into smaller parking places. Some parking places the average driver would think too small to fit a car will be suitable for autonomous vehicles. To find a place to park, the AV can communicate with sensor-based parking places (or even with other vehicles) in a radius several blocks wide; the AV can then go directly to the parking place without having to drive around and search.

Reducing wind drag is another energy-saving feature. Because they can sense the presence of other cars better, AVs can drive closer to one another, which reduces wind drag. This reduction in wind drag would cut fuel usage by an estimated 20 to 30 percent, according to the Rocky Mountain Institute.[277]

### Waste of Money

An American minivan owner who drives 10,000 miles per year spends an average of $8,161 in annual car costs, according to the American Automobile Association (AAA).[278] This is a relatively large sum when you consider that the median wage in America in 2011 was $26,684.[279] Car costs are after-tax expenditures; they absorb more than a third of the average American's income and cost 81.6 cents per mile to drive.

Global monetary losses due to traffic-related injuries cost $518 billion per year; they cost governments between 1 and 3 percent of their gross national product, according to the World Health Organization.[280]

Traffic congestion costs Americans $101 billion in combined delay and fuel costs. That's $713 per year for each commuter.[281] Parking costs American commuters an average of $1,000 per year; crash damages cost an average $1,500 per year.[282]

Autonomous vehicles can pick up and drop off passengers anywhere, relieving us of having to search for and pay for parking. Because they will be better drivers than we are, accidents and their attendant costs will be minimized.

Ultimately, autonomous vehicles will save money by changing the very concept of car ownership. When AVs can pick up and drop off passengers anywhere, most people will choose not to own a car. And those who do own cars may rent them out during the 90 percent of the time they don't actually drive their cars.

# The Accelerating Race to
# Fully Autonomous Cars

Nissan has pledged to have an affordable autonomous vehicle on the market by 2020. "By 2020 we will have a car that behaves like the cars we have on the track here in California, which means that you can sit in the driver's seat, fold your arms, cross your legs, and basically the car will take where you want to go," said Andy Palmer, executive vice president of Nissan.[283] BMW and Mercedes have also pledged to have an autonomous vehicle ready by 2020.

Andy Palmer also said self-driving technology would be available across the entire Nissan portfolio within two vehicle lifecycles after its first self-driving car.[284] He re-iterated that the company was committed to "zero fatality and zero emissions."

How quickly will the market adopt self-driving cars? Acceptance and adoption rates will vary across market segments, demographic groups, and geographies. A study conducted by Cisco Systems found that 57 percent of the drivers it surveyed trust driverless cars.[285] The report found that 95 percent of Brazilians and 86 percent of Indians would ride driverless cars (92 percent of Brazilians would let their kids ride a driverless car!). Only 37 percent of German and 28 percent of Japanese drivers trust self-driving cars. Americans were in the middle of the pack with a 60-percent acceptance rate.

These are surprisingly large numbers for a technology that hasn't hit the market yet. That drivers welcome autonomous-driving cars may speak to the need of many people to find relief from the tedium of sitting in traffic. Sit in Sao Paolo's interminable traffic and you will understand the willingness of the Brazilians to adopt autonomous cars. I recently went to Istanbul, Turkey, where traffic was so bad our taxi driver watched a live soccer game on his iPhone while driving.

The year 2020 may be the first year of fully autonomous cars, but the transition to self-driving cars has already started. To understand how this transition will play out, consider the framework developed by the National Highway Traffic Safety Administration (NHTSA). The NHTSA is the U.S. government agency responsible for developing, setting, and enforcing federal motor vehicle safety standards. The NHTSA has developed a five-level framework to establish clarity when communicating its regulatory efforts to vehicle manufacturers:[286]

Level 0: *No Automation.* At all times, the driver is in complete control of the vehicle and all its primary controls (brake, steering, throttle, and motive power).

Level 1: *One Function-Specific Automated.* One control function is automated.

If multiple functions are automated, they operate independently of each other. The driver has overall control and is solely responsible for the safe operation of the car, but the driver can  cede limited authority over a primary control function that is automated.

Level 2: *Two Functions Automated and Combined.* At least two primary control functions are automated and they work together. By combining different automated functions, the automated vehicle takes control, which allows the driver to disengage from operating the vehicle. The driver can take his or her hands off the steering wheel and foot off pedals at the same time.

Level 3: *Limited Self-Driving Automation.* The driver cedes full control of all critical functions to the car; the driver can then safely engage in activities apart from driving.  The driver is expected to be available for occasional control, but with a sufficiently comfortable transition time.

Level 4: *Full Self-Driving Automation.* The vehicle is designed to drive itself, perform all safety-critical driving functions, and monitor roadway conditions for the entire trip with or without a driver onboard.

Some high-end vehicles already come with software and hardware elements that take driving tasks away from the driver. The 2012 Audi A6, for instance, has sensors, cameras, and software that assist the driver with a number of tasks (see Figure 5.3):
  • Park the car automatically
  • Night vision for pedestrian detection
  • Lane-changing assistant
  • Adaptive cruise control with stop and go

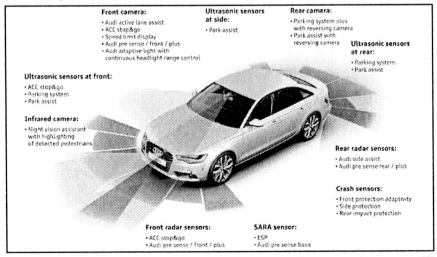

*Figure 5.3 — The 2012 Audi A6 driver assistance systems. (Source: Audi of America)* [287]

Mercedes-Benz "cross traffic assist" technology helps the driver avoid rear-end collisions as well as cross-traffic collisions that occur, for example, at junctions in the road. The car's stereo cameras and system of short-, medium-, and long-range radars generate visual data. The car processes this data to determine whether cross-traffic (from bicycles to trucks) poses a collision risk. If a collision is imminent, the car not only warns the driver, it applies the brakes to bring the car to a full stop.[288]

The BMW X5 offers "traffic jam assistant" technology by which the car drives itself in dense traffic at speeds of up to 25 mph (40 km/h). In other words, the BMW can become a self-driving car in traffic jams.[289]

Add up these tasks and you have a semi-autonomous car. With respect to NHTSA levels there is a continuum from level 1 (single function automation) to level 4 (full automation). Several cars already have traffic jam assistants by which the driver can give control to the car (level 3). Fully automated campus shuttles (level 3) where there is no driver at all (but the car travels only on pre-specified paths) are already in operation.

| NHTSA Level | Example of automation functions | Driver role | Manufacturers offering functions |
|---|---|---|---|
| 0 | None | Driver in complete control | All |
| 1 | -Adaptative cruise control<br>-Emergency dynamic brake assist | Driver in control but can cede it to single function automation | Audi, BMW, Mercedes, Nissan and others |
| 2 | -Adaptative cruise control<br>-Lane-keeping assist<br>-Traffic jam assist | Driver may surf web, read and text must be already to take control when needed | Audi, BMW, Mercedes, Nissan |
| 3 | -Traffic jam assist<br>-City driverless shuttle<br>- Camous driverless shuttle | Driver can sleep | Google, Induct, Nissan |
| 4 | -Full Car Automation | None | None |

Table 5.1 – State of automated vehicles using NHTSA framework. (Sources: Steven Shladover,[290] NHTSA, and author)

Much of the technology that is needed for fully self-driving cars (level 4) is already here. In a matter of a few years, as sensors, computing hardware, and automation software technologies improve exponentially, these high-end features will become cheap enough for lower-priced vehicles to adopt.

## Exponential Technology Cost Improvements

In 2012, Google disclosed that its self-driving car has equipment worth $150,000.[291] At that price Google's car may be too expensive for all but the Ferrari crowd. Many "pundits" questioned whether a self-driving car would be affordable in our lifetime.

Google has not broken down this $150,000 figure except to say that its LIDAR cost $70,000. LIDAR is the rotating, conical, hat-shaped item on the roof of the car (see Figure 5.2). The word LIDAR is a combination of laser and radar. The LIDAR is the technological device that autonomous vehicles use to see ahead and around to all sides.[292]

LIDAR makes up nearly half the cost of a Google car. For the car to drop in price, LIDAR technology will have to drop first. Airborne LIDAR technology, measured by pulse repetition frequency (PRF), has been improving by roughly 100 percent every two years.[293] This represents an improvement rate of 41 percent per year, which is similar to Moore's Law. If this trend continues, LIDAR technology will drop from $70,000 in 2012 to a more manageable $4,481 in 2020 (see Figure 5.4).

Google being an information technology company with a view to the future, I assume that much of the equipment in a Google car consists of computers, communications, sensors, optics, and other advanced technologies. Most of these technologies are improving exponentially. Assuming that Moore's Law applies to all this equipment, the technology components that cost Google $150,000 in 2012 would cost $9,691 by 2020 and drop even further to $3,425 in 2023 (see Table 5.2).

From a cost perspective, it makes sense that Nissan, BMW, and Mercedes-Benz have all announced self-driving cars by 2020.

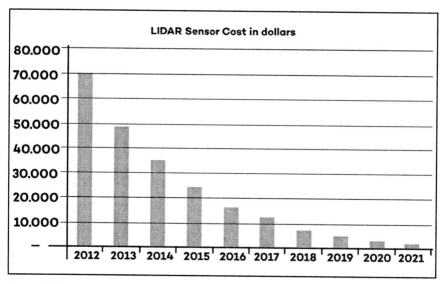

Figure 5.4—Exponentially decreasing cost of LIDAR sensor.

| Year | 2012 | 2013 | 2014 | 2015 | 2016 | 2017 | 2018 | 2019 | 2020 |
|------|------|------|------|------|------|------|------|------|------|
| Cost | $70,000 | $49,645 | $35,209 | $24,971 | $17,710 | $12,560 | $8,908 | $6,318 | $4,481 |

Table 5.2 – LIDAR sensor projected cost.

Technological change can happen faster than anticipated. In late 2013, Google announced that its next-generation self-driving car would use a smaller LIDAR sensor with at least twice the technological performance as before. This LIDAR sensor would cost just $10,000, or one seventh the price of the previous version.[294]

It appears the technology cost curve is accelerating even faster than predicted. This is a point I emphasize in my market disruption class at Stanford: Acceleration is accelerating! I met a Silicon Valley startup CEO who claims to have developed car-quality LIDAR sensor equipment that will sell for $1,000.

LIDAR is just one of several technologies for vehicle visualization. Machines can also use high-definition video technologies to scan and understand the environment. Semiconductor companies are racing to develop computing and sensor hardware and software that can read input from cameras. This technology will make cars more autonomous.

Fujitsu, for example, has announced the "world's first 360° wraparound view system with approaching object detection" (see Figure 5.5).[295] According to Fujitsu, its MB86R24 chip comes equipped with six HD input channels (video cameras) and three display output channels; the chip incorporates "approaching object detection functionality" whereby drivers are notified when people and objects such as bicycles are approaching. Fujitsu's "360° wraparound view system" allows drivers to check their entire surroundings in 3D from any angle. The Fujitsu system costs just 5000 Japanese Yen, about $50, according to a company spokesperson.

*Figure 5.5—Fujitsu MB86R24 "approaching object detection" system." (Source: Fujitsu)*[296]

The cost of technologies that make cars autonomous is decreasing exponentially. By 2020, autonomous vehicle technology will cost no more than rust-proofing and the extended warranties that gasoline car dealers push on car buyers. Meantime, the migration from fully driver-controlled to fully autonomous cars has already started.

## Google, Apple, and Automotive Outsiders

Companies like BMW and Ford are developing vehicle "application programming interfaces" so that third-party software developers can create apps for their cars and drivers can download the apps to their cars from an app store. William Ford Jr., an executive at Ford Motor Company, says that "cars are becoming mobile communication platforms."[297] BMW is already holding "hackathons." The German automaker has a venture capital group (iVentures) to direct investments to software companies.

All this sounds like Silicon Valley getting ready to disrupt yet another technology market — sort of like what Apple and Google did to the mobile phone market. Disruption does not respect industry insiders. In fact, disruption usually comes out of the blue. Tesla's CEO Elon Musk did not come from the automotive industry. Lyndon Rive and Peter Rive did not work in energy before starting SolarCity. Neither Apple nor Google was anywhere near the mobile phone business in the year 2000. The first iPhone was released just seven years ago (June 2007);[298] the first commercial version of Android was released a few months later.[299]

Is it possible that the disruptors of the auto industry will come from outside of transportation?

## Automotive Operating Systems and "Winners Take All" Markets

Software platforms have strong network effects and high switching costs. It's difficult for users of Microsoft Windows software to switch to other software because of the investments they've made, not just in Windows software, but in time and effort to master software skills and complementary Windows technology. Network effects are the reason why Microsoft, despite several major product mistakes (such as Windows Vista and Windows Me), has minted cash for decades with its Windows operating system. It's hard to leave an ecosystem with strong network effects. The barriers to exit are very high. Network effects are also the reason why Apple iOS and Google Android together account for more than 90 percent of the smartphone O/S market.[300]

On the other hand, there are no network effects (and low switching costs) in today's automotive industry. It takes a few minutes to switch your Chrysler SUV for a Ford F150.

It should come as no surprise that Apple and Google want to get into the automotive business. Steve Jobs dreamed of building an iCar. "Look at the car industry; it's a tragedy in America. Who is designing the cars?" Jobs asked, according to Apple Board member Mickey Drexler.[301]

Google is primarily a software company. The Internet giant learned from its acquisition of Motorola Mobility how razor-thin margins are in the hardware business. For that reason, Google isn't likely to get into car manufacturing. More likely, Google will package its autonomous vehicle software in the form of an "autonomous car operating system" and license it to car manufacturers.

Google could develop the automotive equivalent of its Android line of business. It could license its autonomous car software platform to car companies around the world in the same way it licenses its Android O/S to cell phone manufacturers. This would encourage new entrants and commoditize the hardware portion of the car. Because of the strong network effects of operating system platforms, Google developing car software would change the basis of competition in the automotive industry.

A winning software platform for cars might gather an ecosystem of application vendors that would quickly go from 100,000 to a million to one hundred million units and totally disrupt the transportation industry.

Auto giants like GM, Chrysler, and Ford could be the automotive equivalent of former cell phone giants Nokia and Blackberry.

## Car-as-a-Service:
## The Ultimate Disruption Business Model

I have mentioned that the self-driving car is disruptive because, at an individual level, it will save time, energy, and money; on a societal level, it will save time, energy, money, and lives.

Ultimately, autonomous vehicles will be disruptive because they will profoundly change the very nature of car ownership. Cars will go from being an individual object of desire to a revenue-generating business.

Most of us don't really want a car. What we want is mobility on demand. That is, we want the ability to go from point A to point B whenever we want to (or have to). For most people, owning a car is currently the best way to guarantee mobility on demand. However, when the day comes that self-driving cars can pick up and drop off passengers anywhere, anytime, self-driving cars will be preferable to car ownership. What's more, going from place to place in self-driving cars will be substantially cheaper than owning a car.

Again, it's not just technology that is disruptive. It's the business model that the technology enables, namely the car-as-a-service business model.

Imagine you can hail a car anytime from anywhere and have the car show up at your doorstep within minutes. Companies like Zipcar, Uber, and Lyft provide versions of this service today. Imagine being picked up by a self-driving car instead of a human-driven car.

Try this thought exercise: Assume that every vehicle has autonomous technology capabilities and that every car owner makes his or her car available to a company under a car-as-a-service contract. How disruptive is this scenario?

I mentioned previously that car-sharing pioneer Zipcar calculates that each shared car replaces fifteen cars on the road.[302] Assuming that everyone switches to a car-as-a-service model and Zipcar's share-to-own ratio of 15-to-1 applies on a global basis, annual car sales could shrink by a factor of 15. The global car industry sold 82 million units in 2012.[303] If car sales shrink by a factor of 15 due to car-sharing, the industry would only sell 5.5 million vehicles per year, or 6.7 percent of current auto industry production.

Three car companies in the world, Volkswagen, Toyota, and General Motors, sell more than nine million cars each year. Under this scenario, only one of these companies could supply all the autonomous cars in the world — with plenty of spare capacity. (that's assuming none of the big three car companies have been disrupted by the electric vehicle.) Every other car company would have to shut down and go home. Disruption shockwaves would reverberate through the whole auto industry value chain. The oil industry would also shrink.

Furthermore, since the number of cars would drop by more than 93 percent, gasoline demand would also drop dramatically. Autonomous cars would be used more than privately owned cars, so gasoline consumption would not drop by 93 percent. However, autonomous cars are more fuel-efficient, use space better, and don't waste time in traffic or waste time looking for a parking place.

Assuming that each autonomous car consumes three times what a former vehicle did, oil consumption would still drop by 75 to 80 percent! Mathematically speaking, just two countries, Saudi Arabia and Russia, would be able to produce enough petroleum to meet worldwide demand.

The worldwide petroleum and internal combustion engine vehicle would shrink down to just one company selling cars and two countries selling oil!
This scenario assumes that autonomous cars will use internal combustion engines. More likely, the electric vehicle disruption will be well underway. The autonomous vehicle disruption will likely overlap the EV disruption. Think of how the Internet disruption and the cell phone disruption overlapped and complemented each other. They eventually combined to become "mobile Internet."

However, "electric vehicles are the natural platform for autonomous cars," according to Takeshi Mitamura of Nissan's Silicon Valley Research Center. While the electric vehicle disrupts the gasoline car industry, the self-driving car will disrupt it as well and deal the final death blow to whatever remains of the gasoline car industry.

Two industries will be disrupted. The auto industry will shrink massively and the oil industry will either disappear as a supplier to the automotive market (the EV scenario) or shrink massively (autonomous-use-ICE scenario) and eventually disappear altogether (autonomous-EV scenario) as an automotive energy source. Either way, it doesn't look good for the oil industry.

Even if only a relatively cautious share-to-own ratio of 5-to-1 applies after the autonomous car disruption, the automobile market will still severely shrink to at most 20 to 30 million cars per year.

In other words, even assuming that the electric vehicle doesn't disrupt the gasoline car, the autonomous car disruption will cause demand for gasoline to decrease dramatically – maybe by 80 percent.

## Business Model Innovation

Most people think of market disruption in terms of "disruptive technologies." Many times, however, the source of disruption is not a new technology per se but an innovative business model made possible by the new technology.

Consider how Skype disrupted the long-distance telephone market. Many companies had access to the voice over internet protocol (VOIP) technology. It was Skype with its innovative business model that revolutionized the industry.

Car companies have used the same business model for a hundred years. It goes something like this: We make the car, you buy the car, we fix the car until it breaks; repeat every few years. The most radical business model innovation in the auto industry was probably the introduction of car financing by GMAC around 1917. This single innovation helped the car market go from 8 percent of ownership to 80 percent ownership. (See Chapter 2.)

In terms of business models, the auto industry has made very little progress over the last century, but new business models made possible by new technologies are starting to change that. After a car becomes more like a mobile computer-on-wheels, the rules of the game will change dramatically. The automobile will be another product category that falls under Moore's Law. Technology improvements can then happen exponentially. Rather than recall a million cars to repair a defect, auto companies will download new software to the cars over WiFi.

It is not a stretch is to imagine that the auto industry as we know it will not exist in a decade or two. Will electric vehicles disrupt the internal combustion engine auto industry? Will a software platform eat Detroit? It seems to me that several disruption waves are coming: Electric vehicles, software-enabled vehicles and, ultimately autonomous vehicles. Moreover, car sharing will radicalize how we use cars. Combine all this with innovative business models and the gasoline car industry is truly kaput. It's not a matter of if but how and when.

## Disrupting the Car Insurance Industry

Much of the media conversation about autonomous cars revolves around the question of whether the insurance industry will "allow" autonomous cars on the road. This conversation misses the point. Insurance companies should be afraid of the autonomous car because it will disrupt the century-old auto insurance industry.

In Chapter 3, I describe how a company called Climate Corp. uses weather and soil data from the U.S. Government to gather intelligence for its agricultural insurance product. How could a small technology company from Silicon Valley started by two former Google employees with no experience in insurance do this? Data. Lots of data. Big Data.

Said Dan Rimer, a venture capital investor in Climate Corp, "To price its insurance products, Climate Corp's platform ingests weather measurements from 2.5 million locations and forecasts from major climate models, and processes this data along with 150 billion soil observations to generate 10 trillion weather simulation data points, requiring it to manage 50 terabytes of live data at any given time."[304]

The self-driving car is nothing if not a data-generating machine. The Google car gathers more than 1GB of data per second![305] As sensors get cheaper, the amount of data generated by autonomous cars will increase by orders of magnitude. As the number of self-driving cars on the road increases, the amount of data about these cars will grow exponentially.
The company that gathers this data and analyzes it intelligently can price insurance products to a degree that legacy insurance companies can only dream of. This company (let's call it gAuto) can download weather, parking, and energy data from 91,000 openly available datasets maintained by the U.S. government (data.gov). gAuto can also get data from thousands of datasets

maintained by state, county, and city agencies. gAuto will then be able to offer and price insurance products with pinpoint accuracy.

Today, Zipcar profitably rents me a car at a price that includes insurance, fuel, and parking, and Zipcar doesn't have access to the vast amounts of data that a company like gAuto will have. gAuto, with access to vast amounts of data, artificial intelligence, and computing power, will be able to offer a self-driving car service at a price that includes self-insurance, self-charging, and self-parking.

Auto insurance companies that are not getting ready for this are toast. You have been warned.

# Chapter 6:
# The End of Nuclear

*"There is no technical demarcation between the military*

*and civilian reactor and there never was one.*

*- Los Alamos National Laboratory Report LA8969MS,UC-16.*


*"If the world should blow itself up, the last audible voice*

*would be that of an expert saying it can't happen."*

*- Peter Ustinov.*


*"Any intelligent fool can make things bigger, more complex,*

*and more violent. It takes a touch of genius — and a lot of courage —*

*to move in the opposite direction."*

- Albert Einstein.

On April 26, 1986, Chernobyl reactor number four blew up. The explosion caused the biggest industrial catastrophe of the 20th Century. Clouds with four hundred times more radioactive material than that produced by the atomic bombing of Hiroshima blew across Europe and Asia.[306] Radiation levels were so high, they set off alarms at the Forsmark Nuclear Plant in Sweden, located 1,100 Km (660 miles) away from Chernobyl. Soviet leadership and the world learned of the disaster from scientists who measured radioactive clouds in Sweden.[307]

On May 7 and then on May 26, France's "Central Protection Service against Ionizing Radiation" (SCPRI) broadcast its measurements of the radioactive fallout in France. Not to worry, the SCPRI said, the radiation was rather modest, ranging from 500 Becquerels per square meter ($Bq/m^2$) in the eastern part of France to 25 $Bq/m^2$ in the Brittany area, in the country's northwest.[308]
But according to Le Monde, those numbers were not quite right:

> In 2005, a measure from the Institute of Radioprotection and Nuclear Security (IRSN), a successor of SCPRI, that pieced together the fallout from May 1986 showed a far different picture: the deposits of Cesium 137 alone exceeded 20,000 $Bq/m^2$ in certain areas (Alsace, the region around Nice, southern Corsica), with some points in excess of 40,000 $Bq/m^2$.[309]

The radiation numbers that the French government released in 1986 were made up. In France, the real Chernobyl cloud fallout measurements were about a thousand times larger than what the French government had told its citizens.

The real radiation measurement only came to light because the successor organization to the SCPRI was sued in 2001 by the French Association of Thyroid Disease Sufferers (AFMT). This organization accused the government of deliberately falsifying information and failing to take the minimum sanitary measures that neighboring European countries took (for example, banning certain foods).

The government of France knowingly hurt millions of its citizens to protect the nuclear power industry.

After the Fukushima Dai'ichi nuclear disaster in February 2011, the Japanese government similarly misinformed its citizens by minimizing the extent of the damage. Despite an unmitigated disaster that the world had seen in videos and pictures that went viral on the web, despite measurements taken by scientists around the world showing that radiation pollution from Fukushima was on par with Chernobyl, the Japanese government held its own citizens hostage to protect its nuclear industry.

Could the French government nuclear radiation misinformation campaign repeat itself a generation later in Japan? Would it take another two or three decades before Japanese citizens learned the truth about this nuclear disaster?

## Participatory Media, Citizen Science, and the Outing of Nuclear

The disruption brought about by the Internet, cell phone, and personal computer gave citizens the power to create, collect, and publish information. These technologies enabled the rise of an inclusive participatory culture.

Using Twitter, Facebook, and Amazon.com, people participate and contribute data, ideas, and opinions; they don't passively receive content from those in power. Today's technology-enabled participatory culture is the opposite of the closed, secretive, hierarchical culture that characterizes the nuclear power industry.

One week after the March 11, 2011, Fukushima Dai'ichi nuclear meltdowns, a not-for-profit organization called Safecast published its first website and started a sensor network for collecting and sharing radiation measurements. Safecast volunteers soon started taking radiation measurements in Fukushima. Later, they took measurements throughout Japan and, later still, in the rest of the world. Using an open source microcontroller platform called Arduino and radiation Geiger counters from International Alert, Safecast built small, mobile Geiger counters that cost less than $1,000.[310] Safecast called the device "bGeigie," short for "bento Geiger," because it looked like a Japanese bento box. To fund its operations and equipment, the organization raised $35,000 through Kickstarter, the crowd-funding website.

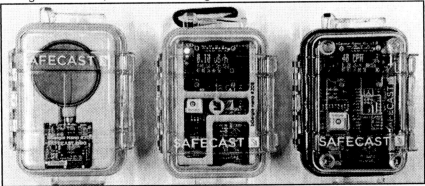

*Figure 6.1—The $450 bGeigie geiger counter kit (about the size of a cell phone). (Source International Medcom Inc)*[311]

Safecast now collects more radiation data at a finer scale than the Japanese government itself. Instead of the government's single Geiger counter per city, Safecast takes nuclear radiation measurements at 50 to 100 meter (150 to 300 feet) resolution every five seconds. It uploads the data every day as zero-license, open-domain content. Anyone can use the data without copyright or financial limitations. On its weather pages, Internet giant Yahoo! Japan has a link that displays radiation information from the Safecast sensor network.

Safecast has uploaded more than 10 million data points, a number that is growing exponentially. It has developed a new version of its Geiger kit sells for $450 to Safecaster volunteers around the globe (see Figure 6.1).[312]

In economics, "regulatory capture" refers to what happens when a state regulatory agency that was created to act in the public interest works instead to advance the commercial or special interests in the industry it is supposed to regulate.[313] In other words, the government agency protects the industry at the expense of the public. Regulatory capture encourages companies to pollute, cut health and safety corners, and take financial and technical risks safe in the knowledge that citizens and taxpayers will bear the costs.

Open data can be considered political when it shines a light on regulatory capture and secrecy. When asked if Safecast is an anti-nuclear organization, co-founder Sean Bonner answered, "Safecast is not anti-nuclear or pro-nuclear; we are pro-data. Data is apolitical."

The nuclear industry and the government agencies that protect it will not be open or transparent anytime soon, but data is seeing the light of day that can bring clarity to the nuclear industry.

And the data are clear: Nuclear is prohibitively expensive, very dangerous, and lethally dirty. Citibank wrote a report about the nuclear industry titled "New Nuclear: The Economics Say No." Nuclear is so prohibitively expensive, the whole industry could not exist at all without regulatory capture to create massive taxpayer subsidies and government protection.

## Regulatory Capture, Decommissioning, and the Prohibitive Cost of Nuclear

In February 2013, Margaret Hodge, a member of the British Parliament, announced that decommissioning the Sellafield nuclear power plant had reached £67.5 billion (US$110 billion).[314] The government was spending £1.6 billion (US$ 2.6 billion) of taxpayer money every year and, she said, "there's no indication of when that cost will stop rising."

When the nuclear industry tells people about the cost of nuclear power it doesn't include the costs of decommissioning (cleaning up) nuclear power plants. Decommissioning is a never-ending source of cash from taxpayers to the nuclear industry.

The Government Committee of Public Accounts, established by the House of Commons, noted in a report that it was unclear how long dealing with hazardous radioactive waste at Sellafield would take or how much it would cost the taxpayer. No nuclear waste has been moved from Sellafield. The nuclear waste is still onsite. According to the Nuclear Decommissioning Authority, the agency in charge of the decommissioning plan, Sellafield should start retrieving hazardous waste in 2015.

Who pays for decommissioning nuclear sites? In nuclear power plants around the world there are two answers: ratepayers and taxpayers. This is not just a British thing. It's a nuclear thing.

In June 2013, soon after the decision to permanently close the San Onofre nuclear power plant in California, its owner and operator, Southern California Edison, started transferring nearly $5 billion dollars in failed repair and decommissioning costs to rate payers. "The traditional way, of course, is all these costs are passed through to the customers, to the rate payers in the form of the rates," said Ted Craver, chairman and CEO of Edison International.[315]

After profiting for decades, nuclear plant operators pack up and turn over the costs of cleaning up their mess to taxpayers and rate payers.

What if energy regulators in the United Kingdom were looking after the citizens they're supposed to serve instead of the nuclear industry? How much solar would the US$110 billion cost of cleaning up Sellafield buy the British people?

To answer that question, assume that regulators allow the creation of a relatively competitive marketplace and that UK solar equipment installation prices reach Australian or German levels. The total installed cost of a 5kW residential system in Australia in July 2013 was US$1.62 (A$1.76) per Watt, according to SolarChoice.[316] At that cost, you could install a total of 67.9 GW of solar PV with $110b (This cost is for unsubsidized solar).

In 2012, electricity demand in the UK was 35.8 GW on average.[317] The peak demand was 57.5 GW. With these figures in mind, the cost to the taxpayer of decommissioning a single nuclear site (Sellafield) would be equal to the total cost of installing unsubsidized solar that would generate 190 percent of the average electricity demand and 117 percent of the peak power capacity in the UK.

Knowing this, you would think that UK regulators would stop any nuclear development and turn to solar (and wind).

## Regulatory Capture, Generation and the Prohibitive Cost of Nuclear

In 2010, a coalition of Conservatives and Liberal Democrats in the UK promised that new nuclear power plants wouldn't receive any taxpayer subsidies.[318] Three years later, the British government backtracked on its promise. It proposed an agreement whereby wholesale nuclear prices would be guaranteed for up to forty years at a potential taxpayer cost of £250bn (US$ 407 billion).[319] The new nuclear capacity to be built would be 16 GW. The British taxpayers would pay up to £15.6 ($25.4) per Watt of nuclear capacity.

Nuclear is already the most expensive method of generating electricity. An unsubsidized peak watt of solar or wind costs less than $2 in Germany and Australia. Why would the British buy into nuclear at more than ten times the capacity cost? The UK is not a sunny place, but Germany, a country with a similar climate to the UK, generates solar power for less than what the British government wants to pay for nuclear generation. Solar has in fact cut the cost of wholesale power by up to 40 percent over the last five years in Germany.[320] Solar is also getting cheaper as nuclear gets more expensive. Why would the UK government want to increase wholesale power prices and hurt consumers to subsidize its nuclear industry?

It gets worse.

The $407 billion in nuclear subsidies would not include the cost of cleaning up and decommissioning nuclear power plants. Consider that, at its peak, Sellafield had just four reactors generating 60 MW each for a total of 240 MW.[321] As mentioned before, Sellafield has already cost £67.5 billion (US$110 billion) to clean up with no end in sight.[322] The 16 GW nuclear expansion that the UK government proposed is 66 times the size of the generation capacity at Sellafield. How many hundreds of billions of dollars will it cost to reprocess the fuel, clean up, and decommission those reactors in forty years?

Furthermore, the $407 billion in nuclear subsidies doesn't include the taxpayer cost of insuring the industry against a nuclear meltdown. The UK is a small country. A Chernobyl or Fukushima-type disaster would have a catastrophic effect on the whole country, cost several trillion dollars, and take countless lives.

So why would a power utility even consider building a nuclear power plant? Three words: government protection and subsidies. Or four words: all gain, no pain.

## Nuclear Subsidies Galore:

## Georgia Power's Vogtle Reactors

When the Atlanta-based Southern Company proposed building its two Vogtle nuclear reactors in 1976, the company said the cost to build both reactors would be $660 million. By the time they were commissioned in the late 1980s, the cost had risen to $8.87 billion — thirteen times the original estimate.[323]

No new ground has been broken on nuclear reactors in the United States since then. Why this drought in new construction? The nuclear industry blames the reactor meltdown at Three Mile Island and what it claims is the public's "irrational fear" of nuclear power. The evidence, however, paints a different picture.

The nuclear industry has been characterized by failure to deliver: cost overruns, delays in construction, and lack of safety. The nuclear industry cannot compete with other methods of energy production.

Nuclear reactors today are about ten times more expensive to build than they were in the early 1970s. The costs keep going up. The nuclear industry may be the only major industry in the world with a negative learning curve. By contrast, solar PV has improved its costs by 154 times 1970. Solar has improved its cost position relative to nuclear by 1,540 times since 1970.

The concept of the learning curve was introduced in 1936 by T.P. Wright of the Aerospace industry. The learning curve says that the more you produce a good or service, the better you become at producing it, so you can make it faster and cheaper.[324] Engineers have measured learning curves for many industries. Learning curves help engineers quantify product cost curves as production scales in the foreseeable future. For instance, if the learning curve in shipbuilding is 20 percent and the first ship costs $100, as production doubles, the next batch of ships will cost $80 ($100 * [1-0.2]). As production doubles again, the next batch will cost $64 ($80 * [1-0.2]), and so on.

You can usually find the learning curves for different industries in engineering books and manuals. The Federation of American Scientists offers an online calculator where you can plug in the learning curve and calculate future costs.[325]

Jonathan Koomey of Stanford University has plotted the actual costs of building nuclear power plants in the United States since 1970 and the results are revealing (see Figure 6.2).[326] As the industry has gained experience building reactors, nuclear power plants have become more expensive.

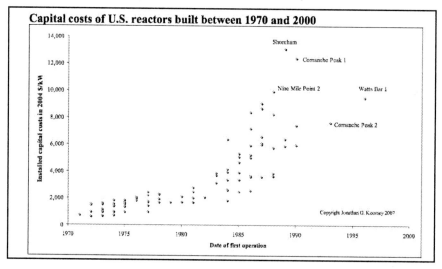

Figure 6.2—Actual capital costs of completed U.S. nuclear reactors between 1970 and 2000. (Copyright: Jonathan G. Koomey 2007)[327]

The time it takes to build nuclear power plants has also lengthened. The industry that promised energy that would be too cheap to meter produces energy that is too expensive to compete. Vermont Law School Professor Mark Cooper has done an in-depth analysis of the nuclear industry in the United States and France. He found, among other things, that the time it takes to build a nuclear power plant has lengthened considerably (see Figure 6.3).[328]

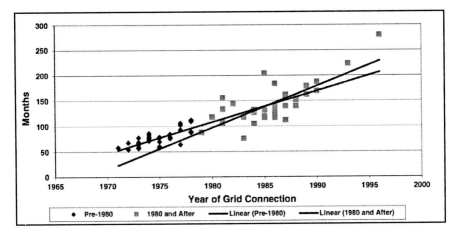

*Figure 6.3—Construction periods in months, pressurized water reactors, by year of grid connections in the U.S. (Source: "Nuclear Safety and Nuclear Economics," Mark Cooper, 2012.)*[329]

According to Professor Cooper's study, the time it takes to build nuclear power plants in the U.S. has risen from about five years in the 1970s to ten to fifteen years in the 1980s. Some reactors have taken more than twenty years to build.

Figures 6.2 and 6.3 show that the cost and construction time to build a nuclear reactor have both risen. The nuclear industry is unique among major industries in that it has what climate expert Joe Romm (a Ph.D. in nuclear physics from MIT) has insightfully called a "negative learning curve." The more experience it gains building its product, the more expensive the product has become and the longer it has taken to build. And the negative learning curve is not a small one. Looking at the data, you can see that, since the 1970s, the industry has increased costs by about ten times and delivery times by about four times.

You don't need a business degree to understand that an industry that keeps increasing costs and missing delivery dates can't survive for long. Especially if its products blow up in your face and literally cost an arm and a leg.

No new nuclear plants broke ground after Vogtle and its peers were commissioned in the 1980s because nuclear power is economically uncompetitive. It survives because the U.S. government gave the energy industry the keys to the taxpayer treasure chest in 2005.

In 2005, the U.S. Congress approved $18.5 billion in new loan guarantees for the nuclear industry, according to the Nuclear Energy Institute, an industry lobby group.[330] The 2005 Energy Act authorized the U.S. Department of Energy to guarantee up to 80 percent of nuclear project costs. It also provided extra insurance of $2 billion for cost overruns and another $1 billion in production tax credits spread over the first eight years of a reactor's life.[331] After the 2005 Energy Act, the industry loudly predicted a nuclear rebirth. Hoping to emulate the high market penetration of the nuclear industry in France, the nuclear industry borrowed a French word (renaissance) to describe its rebirth and promptly went to work.

In 2006, Georgia Power, a Southern Company subsidiary, announced it would build two new 1.1 GW reactors at Vogtle, the Vogtle 3&4. The U.S. Nuclear Regulatory Commission gave its approval and the project broke ground in April 2009. Georgia Power estimated the cost to build the two reactors at $14 billion. They were to start operation in 2016 and 2017 respectively.

In 2009, the Georgia Senate approved Senate Bill 31, which allows Georgia Power to collect up to $2 billion from rate payers to finance the new Vogtle nuclear plants.[332] Rate payers would finance the nuclear power plants while they were being built.

On February 6, 2010, the Obama Administration offered a federal loan guarantee of $8.33 billion for the construction of the reactors.[333] Construction of the Vogtle 3 reactor officially began on March 12, 2013 with the pouring of concrete for the nuclear island.

The story of Georgia Power and its Vogtle 3&4 nuclear plants illustrates how the nuclear industry uses taxpayer money to finance nuclear power plants. Look how Georgia Power took advantage of its cozy relationship with regulators and policy-makers:
- Federal loan guarantee: $8.3 billion.
- Financing from rate payers: $2 billion.
- Production tax credit: $1 billion.

Assuming that the Vogtle 3&4 nuclear plants come in under budget and actually produce energy, $11.3 billion of the $14 billion cost of building the plants will have come directly from taxpayers.

But don't expect these projects to be finished on time or under budget. Remember that the original Vogtle 1&2 reactors were thirteen times over budget. Vogtle 3&4 have barely broken ground and are already two years late and up to two billion dollars over budget. The project is now expected to cost up to $16.5 billion and open in 2018 and 2019.[334]

Is anyone surprised that a nuclear project is late and over budget? The industry is pathological about over-promising and under-delivering. Every nuclear plant built in the United States has been late and over budget, or else has been canceled.

According to the Congressional Budget Office, the 75 nuclear plants built between 1966 and 1986 were three times more expensive than their builders originally estimated.[335]

Moreover, of the 253 nuclear plants that were originally ordered between 1953 and 2008, 121 (48 percent) were canceled before completion.[336] Of the 132 plants that were built, 21 were permanently shut down due to reliability or cost problems, while another 27% completely failed for a year or more at least once, according to Rocky Mountain Institute energy expert Amory Lovins.[337] "Many nuclear plants operate profitably now because they were sold to current operators for less than their actual cost."[338]

The Washington Public Power Supply System Service (WPPSS), now Northwest Energy, ordered five nuclear power plants in the early 1970s. Delays and cost overruns caused the utility to cancel two of five, halt construction on two mores, and default on $2.25 billion, then the largest municipal bond default in history.[339] Only one of the original five nuclear plants, the Columbia Generation Station, is working today.[340]

A Forbes magazine cover story from February 11, 1985 titled "Nuclear Follies" said, "The failure of the U.S. nuclear power program ranks as the largest managerial disaster in business history, a disaster on a monumental scale ... only the biased can now think that the money has been well spent. It is a defeat for the U.S. consumer and for the competitiveness of U.S. industry."

Is Georgia Power shamed or worried about cost overruns? Not at all. The 2005 Energy Act conveniently set aside $2 billion of taxpayer money to account for cost over runs.
For Vogtle, the treasure chest of taxpayer subsidies keep adding up:
- Federal loan guarantee: $8.3 billion.
- Financing from rate payers: $2 billion.
- Production tax credit: $1 billion.
- Cost-overrun protection: $2 billion.

Forgive me if all this reminds me of Wall Street's "Mickey Mouse" loans that brought about the Great Crash of 2008 and pulled the world economy down with it.

Nuclear power is an industry that has proven that it can deliver an ever more expensive product and take longer to produce it. Nuclear is already an uncompetitive industry with a negative learning curve. If energy were based on market forces, the nuclear industry would have been out of business long ago.

The only way nuclear can stay afloat is to take subsidies from the taxpayer. Sadly, some governments are delivering on those subsidies. The Obama

Administration's 2012 budget called for tripling the nuclear loan guarantee program from $18.5 billion to $54.5 billion.[341]

## Insuring the Uninsurable:
## Taxpayers Bailing out Nuclear

As expensive and as painful as government protection and subsidies to the nuclear industry are, the most expensive subsidy may be yet to come: nuclear insurance. In the U.S., nuclear insurance is called the Price-Anderson Nuclear Industries Indemnity Act. Whatever the failings of the nuclear industry have been so far, bailing out — in other words, insuring — the nuclear industry is one area where failure could bankrupt not just a company or an industry, but an entire nation.

In 2009, there was one accident for every 1.4 million Western-built jet aircraft flights, according to the International Air Transport Association.[342] Based on those figures, you have a 0.00007 percent chance of being in an aircraft accident. On the other hand, 1.5 percent of all nuclear reactors ever built have melted down, according to Stanford Professor Marc Jacobson.[343] The likelihood of a nuclear plant reactor having a meltdown is almost 1 million times greater than the chances of your next aircraft flight crashing.

Now imagine you were told that the flight you and your family are about to board has a 1.5-percent chance of blowing up. Would you get on board?

The Fukushima nuclear disaster has reminded us once again that nuclear power is not safe.

The Fukushima nuclear disaster was a tragedy of proportions not seen since the Chernobyl nuclear disaster. The Japanese government has not been forthcoming about the human, environmental, and financial cost of this tragedy. On the contrary, Japanese Prime Minister Shinzo Abe rushed through sweeping legislation that prevented the public from accessing any information about Fukushima that any bureaucrat deemed a "state secret." "This may very well be Abe's true intention — cover-up of mistaken state actions regarding the Fukushima disaster and/or the necessity of nuclear power," said political science professor Koichi Nakano of Sophia University.[344]

After decades of government assurances that nuclear was safe, clean and cheap, Japanese citizens are paying for this disaster not just with their lives and their health, but also with their wallets. Japanese taxpayers have learned the hard way what happens when a nation insures the nuclear industry against any and all major disasters.

But the fact that taxpayers insure the nuclear industry is not a Japanese thing. It's a nuclear regulatory thing. If there is nuclear power plant in your country, then you, too, are in the nuclear insurance business. Do you know how much your liability is?

In the United States, Congress has decided that the taxpayer is liable for nuclear disasters. It's the law of the land. It's called the Price-Anderson Nuclear Industries Indemnity Act.

Congress passed the Price-Anderson Act in 1957 in an effort to protect the nascent civilian nuclear industry. In 1957, the private insurance industry did not have enough data to accurately price insurance premiums for a nuclear power plant. But the nuclear industry has grown and matured since 1957. Today, more than 430 reactors in 31 countries provide 370 GW of nuclear power capacity, according to the World Nuclear Association.[345] The nuclear industry has achieved high penetration in countries such as France, Japan, Russia, and the United States. France has 59 reactors that generate about 75 percent of the country's electricity.[346] Before the Fukushima disaster, Japan had 50 reactors that generated about 30 percent of the country's electricity.[347] The U.S. has about 100 reactors that produce about 19 percent of the country's electricity.[348]

The world has been building, operating, maintaining, and generating nuclear energy for six decades. Private insurers have enough data to quantify the safety of nuclear plants. They have enough data to create an insurance product for nuclear power plants, right? Yes, they have enough data but no, they won't insure nuclear. Not a single insurance company has stepped forward to cover the full costs of a nuclear disaster. Private insurance companies insure buildings such as the new Freedom Tower (which was built after the World Trade Center terrorist attacks). Private insurance companies insure against the risk of hurricanes and airplane accidents. But no private insurance company will insure the full costs of nuclear.

Assume for a moment that there was a market for nuclear power insurance like there is a market for, say, car insurance or solar power plant insurance. What premium would insurance companies charge to insure nuclear power plants?

The German government (where taxpayers also insure the nuclear industry) commissioned a study to answer that question. The April 2011 report concluded that, for a private insurance company to insure a nuclear plant, the insurance premium would be in the 0.139 €/kWh (19.9 ¢/kWh) to 2.36 €/kWh ($3.39 ¢/kWh) range.[349]

To put those figures in context, the city of Palo Alto has a 25-year power purchase agreement in which it pays 6.9 ¢/kWh for solar.[350] So the total price that Palo Alto pays for each unit of solar energy (6.9 ¢/kWh) is about one third of the lower estimate of the insurance premium that a nuclear producer would have to pay for each unit of energy (19.9 ¢/kWh).

Here's another way to look at it: The solar independent power producer (IPP) that supplies Palo Alto has to pay capital costs, installation costs, management costs, insurance costs, operation and maintenance costs, taxes, permit costs, and other costs. After all that it has to make a small profit from the energy it sells to Palo Alto for 6.9 ¢/kWh. All those costs put together, plus the developer profits, are about one third of the insurance premium that a nuclear power plant would have to pay for each energy unit (kWh). And that's for the low-end nuclear insurance premium estimate (19.9 ¢/kWh). If you take the high-end estimate ($3.39 ¢/kWh), the total cost of solar electricity is fifty times lower than the premium to insure nuclear.

Nuclear is uninsurable. When utilities and regulators talk about the "cost of nuclear," they don't include insurance subsidies from taxpayers. The costs of a disaster such as Fukushima or Chernobyl are not tabulated on spreadsheets when the "cost of nuclear" is calculated.

No wonder the nuclear industry wants to freeload on taxpayer insurance. Nuclear is not economically viable as-is. Having to pay for insurance (if it were available in the marketplace) would put nuclear out of business immediately and irrevocably.

The German report also concluded that the expected damage value of a nuclear disaster would be €5.756 trillion Euro ($8.27 trillion). Germany's gross domestic product (GDP) in 2012 was $3.4 trillion, according to the World Bank. The cost of a nuclear disaster in Germany, according to the report, would be about 2.4 times the size of the German economy.[351] In order words, a single nuclear disaster could bankrupt the largest economy in Europe and the fifth largest economy in the world.

Nuclear is not just prohibitively expensive, it could also bankrupt a whole country. And the smaller the country's economy, the swifter the country would collapse. Russia's GDP is now about $2 trillion, but in the late 1980s, it was closer to $500 billion.[352] Michael Gorbachev said that the 1986 Chernobyl nuclear disaster "was perhaps the real reason the Soviet Union collapsed five years later."[353]

Moreover, nuclear disasters can't be contained to their immediate surroundings. "We initially believed that the main impact of the explosion would be in Ukraine, but Belarus, to the northwest, was hit even worse, and then Poland and Sweden suffered the consequences," said Gorbachev.

# The Nuclear Death Spiral

Faced with such daunting data, Germans decided after the Fukushima disaster to shut down eight nuclear reactors immediately and shut down their whole nuclear industry by 2022. Meanwhile, they accelerated what already was the most ambitious clean energy program in the world, a program based on solar and wind generation, energy efficiency, and electric vehicles.

Most countries in Europe have also accelerated the nuclear phase-out process that they started after the Chernobyl nuclear disaster in 1986. Italy held a referendum in June 2011 in which 95 percent of voters resoundingly rejected their prime minister's push for a new nuclear energy program.[354] Global Data expects 150 of Europe's 186 nuclear power plants (80 percent) to shut down by 2030.[355] All of Japan's fifty nuclear plants have shut down. Japanese citizens gathered eight million signatures against the government's plan to restart its nuclear power plants.[356] Despite massive public opposition to nuclear, the Japanese government may reopen a few reactors. However, the Japanese nuclear program is clearly on its deathbed.

In the United States, where the Obama Administration has tripled subsidies for new nuclear power plants and even referred to nuclear as "clean energy" (repeating the industry's clever attempt at positioning itself as clean), nuclear power plants are shutting down at a pace not seen in decades. In April 2011, NRG Energy CEO announced that, for "financial reasons," his company would stop work on two new reactors being built in south Texas.[357] Four more nuclear power plants were closed in 2013: Vermont Yankee and Wisconsin Kewaunee because they could not compete in wholesale markets; Crystal River in Florida because of structural damage; and San Onofre in California for a number of reasons, including structural damage, failed repairs, and safety concerns.[358]

The existing fleet of nuclear power plants in the U.S. is getting older, more inefficient, more expensive to operate and maintain, and increasingly less competitive. A report by investment bank Credit Suisse points out that the number of nuclear plant outage days has been increasing significantly (see Figure 6.4). This increase in outage days has increased the costs for repairs and upgrades. The San Onofre nuclear plant, for example, closed in 2011 so its malfunctioning steam generators could be replaced. After spending $670 to make the repairs, the steam generators were deemed unrepairable. San Onofre has not reopened and is scheduled to be decommissioned.

Operations and maintenance (O&M) costs have gone up by 4.8 percent annually; fuel costs rose by 9.1 percent per year during the years 2007–2011; the fully loaded costs are expected to continue to rise by 5 percent per year in the foreseeable future, according to a Credit Suisse report.[359]

An analysis by Mark Cooper of the Vermont Law School lists 38 nuclear plants that are "at risk of closure" for purely economic reasons, of which ten face "particularly intense challenges."[360] Nearly all of these plants are among the 47 nuclear power plants that have to compete daily in the open wholesale markets in the U.S. [361]

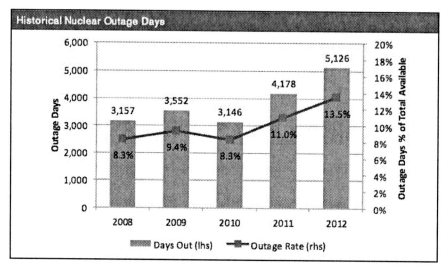

*Figure 6.4—Historical outage days, U.S. nuclear industry (Source: Credit Suisse)* [362]

According to these analyses, nearly half of existing U.S. nuclear power plants may have to close soon purely for economic reasons.

Even plants in nuclear-friendly state regulatory environments face another reason for their demise: the flattening of peak prices for electricity due to the higher penetration of solar. Peak power prices are many times higher than average prices; they provide nuclear power plants with high margin opportunities. But peak prices happen to coincide with hot sunny days when solar generation is at its highest. Can nuclear make money in the off-peak market? "Nuclear plants will struggle to make positive cash margins in off-peak hours," according to Credit Suisse.[363]

French nuclear giant Electricité de France (EDF), the world's largest operator of nuclear plants, decided to exit the U.S. market.[364] EDF invested six years and $2.7 billion in a nuclear "renaissance" that no one expects anymore. EDF Chief Financial Officer Thomas Piquemal said that this was the "last chapter of their U.S. adventure."

On December 16, 2013, USEC (formerly the Uranium Enrichment Corporation), the only American uranium enrichment company in the business, announced it would file for bankruptcy.[365] USEC had received $257 million in aid from the U.S. Department of Energy during the previous two years for its American Centrifuge project, a centrifuge uranium enrichment technology for fueling nuclear power plants. The project was supposed to be completed by 2005 and cost $1.7 billion. Instead, it is expected to cost $6.5 billion and not be finished until 2016. A month prior to its bankruptcy announcement the company declared, "At current market prices, we do not believe that our plans for (the American Centrifuge Plant) commercialization are economically viable without additional government support."[366] Is anyone surprised when a nuclear project is not financially viable, is at least twelve years late, and costs almost four times its original estimate?

As solar and wind increase penetration nationally, nuclear is toast. New nuclear plants like Vogtle, assuming they will eventually be built and commissioned, will not be able to compete in an open market. These "new nuclear power plants" are expected to produce electricity at an uncompetitive cost between 25 ¢/kWh and 30 ¢/kWh.[367] The average retail price of electricity to residential customers in the U.S. in September 2013 was 12.5 ¢/kWh.[368] So the expected cost of "new nuclear," when and if the new plants ever get built, would be twice the retail price of today's retail electricity. Add the cost of transmission, distribution and utility overhead, and nuclear could sell for three times today's retail prices. By contrast, solar costs are going down dramatically. First Solar's 50 MW Macho Springs project will sell solar to El Paso Electric for 5.79¢/kWh.[369] That's about five times less than "new nuclear."

When the Chicago-based utility Exelon announced it was scrapping its nuclear reactor project in Victoria County, Texas, the company said it was all about the economics: "...economic and market conditions make construction of new merchant nuclear power plants uneconomical now and for the foreseeable future," the company said.[370] John Roe, CEO of Exelon, a company with a portfolio at the time that was 93-percent nuclear, dismissed the so-called renaissance altogether. "Don't kid yourself that [nuclear] is economic. Building out nuclear capacity would require as much as $300 billion in federal loan guarantees and other subsidies."[371]

When EDF packed up and left the U.S. market, the company they went straight to London where the UK government has promised, for a period of 35 years, to guarantee the nuclear industry a price that is twice the current wholesale price for electricity.[372] The European Commission calculated how much taxpayers would have to subsidize a nuclear plant to be built by EDF at Hinkley Point; the amount was £17 billion (US$27.8 billion). This amount does not include decommissioning, cleaning up, and insuring the Hinkley Point reactors. EDF said without this "reform" (taxpayer subsidies), the investment in Hinkley Point will not happen. Günther Oettinger, the European Union's energy commissioner, described the EDF nuclear project in the UK as "Soviet" in style.

The shrinking nuclear capacity means that nuclear is going into a death spiral. The vaunted nuclear rebirth never happened. What we got instead was the nuclear re-death.

## Disrupting the Nuclear Zombie

Nuclear is a zombie — not quite alive but not dead either. The real danger from zombies is that they want to suck the life out of the living. Nuclear depends on taxpayer subsidies and always has. The nuclear lobby has succeeded in positioning nuclear as a "clean" alternative to fossil fuels and getting even more subsidies than before. You can fool most of the people most of the time, but the lethal realities of Fukushima and the brutal market reality of an uncompetitive nuclear industry are far too difficult to hide in open societies and open energy markets.

However, in the course of building a large market in the U.S. and Europe, a critical mass of engineers, academics, and suppliers coalesced around the nuclear industry. As the industry shrinks, these engineers and academics will leave for better jobs elsewhere; new university talent will not seek careers in a dying industry. Suppliers will go out of business or shift to other industries. Having fewer suppliers focused on nuclear will mean even higher costs, longer delays, and the loss of whatever economies of scale were there before. The R&D will have to be spread over fewer reactors, which will raise the per-unit costs even more.

The end result will be an industry with prohibitively high costs having to increase those costs even more. The industry will also be more dependent on higher and higher taxpayer subsidies and government protections. A once attractive industry will never again attract top science and engineering talent, which means fewer technology breakthroughs and more quality problems, which in the nuclear industry means more safety issues — and the resulting accidents.

The nuclear industry has entered a vicious cycle of market death.

Solar and wind keep growing their market share, increasing quality, and cutting costs. As solar and wind decrease costs and beat nuclear in the retail and wholesale electricity markets on a daily (and nightly) basis, more nuclear plants will have to shut down for purely economic reasons.

I mentioned earlier in this chapter that solar has improved its cost position relative to nuclear by 1,540 times. Solar costs are expected to drop another two thirds by 2020. Even if nuclear costs don't go up anymore (an unlikely scenario), solar will have improved its cost position relative to nuclear by

4,620 times. For all the reasons stated here, the more likely scenario is that new nuclear costs keep rising and, by 2020, solar will have improved its cost position by six-thousand or more times relative to nuclear.

Like any disrupted industry, the nuclear industry's death spiral will end quickly (albeit not painlessly). The United States will "need only a handful of nukes that we'll need to keep running as baseload plants," said David Crane, CEO of NRG Energy. That's a 95-percent drop in the number of nuclear plants in the U.S.

Instead of putting money into the nuclear black hole, NRG is investing in solar. The company is developing solar in large-scale projects such as the 377 MW Ivanpah concentrating solar power tower plant in California's Mojave Desert. The company is also investing in distributed solar PV projects around the country.[373]

Baseload solar is here. We may not even need David Crane's "handful of nukes." Kevin Smith, CEO of SolarReserve, told my class at Stanford University that baseload solar is already cheaper than a new nuclear plant. Located halfway between Las Vegas and Reno, SolarReserve's new power plant in Crescent Dunes, Nevada, has ten hours of energy storage, so it can sell solar electricity on demand whenever customers need it. SolarReserve has a 25-year contract with NV Energy to sell power at peak demand hours for 13.5 ¢/kWh. SolarReserve's 110MW plant is a first of a kind in the U.S. Kevin Smith expects to cut costs by half over the next few years. Starting in 2014, Nevada Power Company will light up the Las Vegas Strip with solar power. What happens in Vegas will be powered by solar!

The end of nuclear will mean the end of a popular deception — that the "civilian nuclear" industry as a viable industry. We will have to pay to clean up the nuclear mistake for generations to come in places like Sellafield, Chernobyl, and Fukushima. But make no mistake, nuclear is already obsolete. The nuclear industry is imploding because it's too expensive, too dangerous, and too dirty. Let this zombie go before it does more irreversible damage to the living.

# Chapter 7:
# The End of Oil

*"When the wind of change blows, some build walls,*

*others build windmills."*

*- Chinese Proverb.*

*"I'd put my money on the sun and solar energy.*

*What a source of power! I hope we don't have to wait*

*until oil and coal run out before we tackle that.*

*- Thomas Alva Edison, 1931.*

*"The Stone Age didn't end for lack of stone,*

*and the oil age will end long before the world runs out of oil."*

- Sheik Ahmed Yaki Zamani, former Saudi Arabia Oil Minister.

On May 11, 2011, Saudi Arabia announced that it would embark on a twenty-year, $109-billion project to deploy 41,000 MW of solar power (the peak equivalent of 41 nuclear power plants).[374] The kingdom burns 523,000 barrels of oil per day to generate electricity and desalinate water. By 2030, as its population, economic activity, and energy demand grow, Saudi Arabia may burn 850 million barrels of oil a year, or 30 percent of its crude output, to generate electricity, according to Abdullah Al-Shehri of the kingdom's Electricity & Co-Generation Regulatory Authority.[375]

The economics of Saudi Arabia's massive solar electrification project make sense. The cost of generating electricity and desalinating water with solar technologies is a small fraction (10 to 20 percent) of what it costs using oil energy. Instead of burning oil that can be sold for $100 or more per barrel on the open market, the Saudis will use solar, which produces electricity at a small fraction of the cost, the equivalent of less than $20 per barrel.

The stone age did not end because we ran out of stones. It ended because rocks were disrupted by a superior technology: bronze. Similarly, the oil age will not end because we run out of oil. It will end because oil will be disrupted by superior technologies — solar, electric vehicles, and autonomous cars — and the business models that they enable.

Saudi Arabia, the largest oil producer in the world, has seen the light and is showing the way out of the oil age.

## Solar Exponential Cost Improvement
## Relative to Oil

Saudi Arabia had better hurry up. As enlightened as its plan to produce 41GW of solar is, it may not have another twenty years of oil income to fund its solar plans. Solar has cut its costs exponentially relative to oil; the solar cost advantage will only improve over the next few years.

Imagine a world where oil prices followed the same cost curve as solar. In 1970, oil cost $3.18 per barrel[376] and gasoline retailed for $0.36 per gallon in the U.S.[377] If oil had dropped in price at the same rate as solar, a barrel of oil would cost about 2 cents and a gallon of gasoline would cost $0.00234 per gallon. Four gallons of gasoline would cost about one cent! In this imaginary universe, you could fill up a 15-gallon tank for 3.5 cents. Instead, oil hovers around $110 per barrel and filling up a gas tank costs more than $50.

In reality, solar PV is 154 times cheaper now than it was in 1970 (costs went from $100/W down to 65 ¢/W), while oil is 35 times more expensive (oil went from $3.18/barrel to $110/barrel) (see Figure 7.1). Put these numbers together and you find that solar has improved its cost basis by 5,355 times relative to oil since 1970.

If you think that an industry can compete with a technology that has improved its cost position relative to yours by more than five thousand times, clearly you're in denial about the impending disruption.

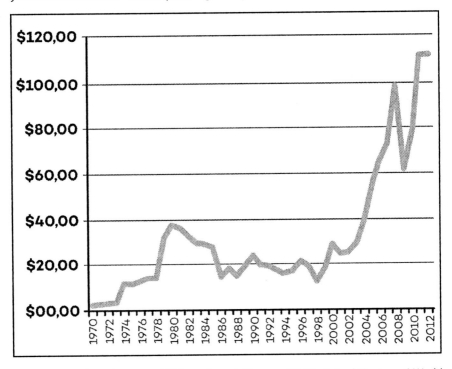

*Figure 7.1—Oil crude prices, US$ per barrel (Source: BP Statistical Review of World Energy.)*[378]

In an industry that is about to be disrupted, a company has three choices:
1. Get out. That is, sell at high prices while you can.
2. Invest in the disrupting industry.
3. Die.

The Kingdom of Saudi Arabia has chosen option 1 and 2. The Saudis are selling oil at high market prices while they invest in a new energy infrastructure based on solar, the disrupting technology.

It's going to get worse for oil. Solar PV costs are projected to drop by another two thirds by 2020. Assuming that oil remains at $110 per barrel, this means that solar will have improved its cost position relative to oil by about 12,000 times.

Oil is a commodity with geo-political dimensions, which makes any future price prediction tenuous at best. Whatever scenario you believe in, oil is in trouble relative to solar. Whether you think oil is going to fall back to $55/bbl or rise to $220/bbl by 2020, PV will have improved its cost curve by either six thousand or twenty-four thousand times. Either way, oil will be disrupted.

If you think that solar doesn't compete with oil, think again. Oil will be disrupted by three complementary disruptive waves:
1. The electric vehicle will disrupt the internal combustion engine vehicle. The EV will make gasoline and diesel obsolete for transportation (see Chapter 4).
2. The autonomous car will make transportation ultra-efficient, decreasing waste and shrinking the fleet of cars around the globe probably by an order of magnitude (see Chapter 5).
3. Solar PV will displace petroleum as a source of energy in power generation (diesel) and heat and lighting (kerosene), both of which still provide expensive energy to billions of people around the world. Solar is already cheaper than diesel and kerosene.

## The End of the Canadian Oil Sands

Some of the world's most environmentally destructive oil projects are also financially expensive. The Canadian Oil Sands project will soon be a stranded asset not because the project is environmentally destructive but because it is not financially viable.

To sustain the capital investments needed for offshore drilling or sand oil projects, the market price for oil must be consistently high. For example, it costs between $65 and $100 per barrel to produce oil in the Canadian Oil Sands.[379] The actual price per barrel depends on the type of project and the technology used. For new steam-driven projects, the breakeven point (the minimum market price investors require to make a cash profit) is $65–$70; for mining projects the breakeven point is closer to $90–$100, according to Wood McKensie, a consulting firm. Alberta's oil sands will need about $650 billion in capital investments over the next decade, according to Canada's Natural Resources Minister, Joe Oliver.[380]

Investors want the highest possible return on their capital investments. If they don't believe the price of oil will stay consistently above the breakeven point, it is not rational for them to invest. If investors expect oil prices to be in the $50/bbl range (for the foreseeable future), the Canadian Oil Sands will not be developed at all for lack of investors. If investors expect oil prices to float around $80/bbl, the steam-driven projects may be developed because their cost per barrel is in the $65–$70 range, but the more expensive mining projects will not be developed.

A similar logic applies to offshore oil development. Figure 7.2 shows the dramatic difference in the cost per barrel to develop a barrel of oil onshore and offshore.

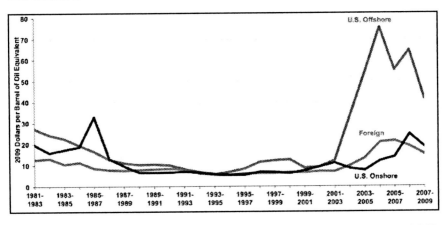

*Figure 7.2—Development costs for FRS companies from 1981–1983 to 2007–2009. (Source: EIA)*[381]

Notice in Figure 7.2 that the cost per BOE (barrel of oil equivalent) peaked around $80 and then dropped below $50 in the 2007–2009 period. This spike in the cost per BOE occurred right before the BP Gulf Oil disaster of April, 2010. Were oil companies improving technology costs or were they just cutting costs without regard for worker safety and environmental consequences?

The Brent benchmark is probably the best-known benchmark for global crude oil market prices. The Brent crude oil benchmark has been above $100 since 2011.[382] The West Texas Intermediate (WTI) benchmark has clocked in slightly below $100 during the same time period. If investors believe that oil prices will remain at or above the $100 range, most of the Canadian Oil Sands projects may be developed after all.

However, oil prices are extremely volatile. Per barrel oil prices of $100 have become commonplace, but a quick look at Figure 7.3 shows that oil prices were actually quite low during the 1990s. Oil was around $10 per barrel as recently as February 1999.

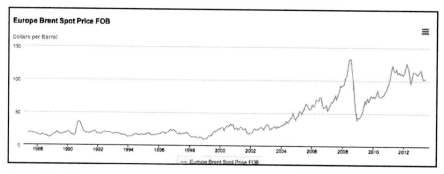

*Figure 7.3—Brent oil spot prices 1987-2012. (Source: EIA)*[383]

Should oil demand shrink — say by 2030 because electric and autonomous vehicles disrupt the auto industry and solar disrupts diesel — oil prices will have nowhere to go but down.

Some in the fossil fuel industry argue that high market prices are needed as an incentive to exploration and investment. However, as Figure 7.2 shows, until the year 2000 prices were relatively low and plenty of oil was nevertheless being developed. Why? Notice in Figure 7.1 that the costs to develop oil onshore were actually quite low, mostly in the $20 range. The fact is oil companies are mostly lifting oil at $20 and selling it at $100. That's five times production costs. Not a bad gig if you can get it.

When oil prices return to, say, 1990s levels in the $20-$30 per barrel range, the following will happen:

1. Only low-cost producers will survive. Only highly productive fields with breakeven points below $15-$20 will be developed.

2. The majority of environmentally catastrophic developments (which happen to be massively expensive) will be stranded. Canadian Oil Sands with its $65/bbl costs, offshore developments with their $60-$70 breakeven points, and deep Arctic developments with as yet unknown financial and environmental costs will be put on hold. Rational investors will never come back to these projects. Ever.

Prime Minister Stephen Harper of Canada once said that developing the Canadian Oil Sands was an epic undertaking on the scale of the Egyptian pyramids, only larger. He was right but not in the way he intended. The Canadian Oil Sands investments will soon be stranded assets of monumental size. Like the pyramids, they will be a massive tombstone and a reminder for future generations of the damage that megalomaniacal investments can bring to a formerly wealthy and powerful society.

## The World's First Solar Nation

On October 29, 2012, the South Pacific island-nation of Tokelau became the first nation in the world to go 100-percent solar. Tokelau is a decidedly small nation with a population of 1,411 people spread over 12 km$^2$ on three atolls. Tokelau switched to solar because the nation had a problem that is typical of diesel-powered economies.

Tokelau used to spend about NZ$ 1 (US$ 0.83) million on diesel fuel per year.[384] This doesn't sound like much until you consider the island nation's total GDP, US$1.5 million.[385] Fifty-five percent of the island's income was spent for diesel fuel to power its economy.

This all changed when the island went solar. Much of the conversation about electricity focuses on the levelized cost of energy. Few people take into account the quality of life that dependence on the fossil fuel drug engenders. Diesel is so expensive, diesel generators in Tokelau were shut off at night to save money. Medicines that needed to stay cool round-the-clock would go bad; patients who needed treatment at night would go untreated.[386] Diesel was delivered by boat once a month. Sometimes the islands ran out of diesel altogether and would have to wait for the next shipment in the dark. Diesel was at best an intermittent source of energy.

Now that they are off diesel, Tokelauans will nearly double their income, raise their quality of life, and have a 24/7 supply of energy.

PowerSmart Solar, a company based in Tauranga, New Zealand, built Tokelau's solar installations. Dean Parchomchuck, a co-founder of the company, led the installation team on the islands. I met him over coffee at the University of Auckland Business School, where he told me that his company's solar installation took a grand total of 22 weeks. The first atoll took ten weeks and the second and third atolls took about six weeks each. Here you can see the solar installation learning curve at work. Had there been a fourth atoll, the installation time there would have been even shorter. The three solar power installations consist of 1.5 MW of solar photovaltic panels and on-site battery banks that store the energy for nighttime and rainy day usage.

The solar power plant cost NZ$ 7.5 million (US$ 6.2 million) and was financed by a grant from New Zealand Aid Programme.

It took an island nation in the South Pacific just five months to go from all-diesel to all-solar. It's a small energy market but the lesson is clear: When disruption happens, it can be swift.

## The End of Diesel Is the End of Energy Poverty

In 1991, India had five million phones. In May 2012, the country had 960 million phones and was adding eight million new phones each month![387]

In 1991, India's leadership  passed new legislation to break up the old monolithic, centralized, and inefficient telecom industry. The goal was to provide telecommunications for all. In less than two decades India's phone usage went from about 0.05 percent to 80 percent of the population, a stunning 19,100-percent growth. India now has the second largest telecom market in the world.

When India's power grid collapsed in August 2012, more than a billion people were without access to grid power. Six-hundred million were affected by the blackout. However, another five- or six-hundred million did not even notice the blackout because they didn't get their energy from the grid.[388] This population relied on diesel, kerosene, or firewood for their energy needs. Indians pay up to $2 per kWh for this energy, more than ten times the cost of unsubsidized solar. Not having access to reliable and inexpensive power can keep whole populations in a never-ending poverty cycle.

The Indian government itself is helping to exacerbate the fuel problem. In 2011, India's fossil fuel subsidies amounted to $39.7 billion, according to the International Energy Agency.[389] India itself is paying billions to help perpetuate the poverty cycle of its own citizens.

India is not alone. "One of the energy industry's best-kept dirty little secrets is the $300 to $400 billion per year world government subsidy of diesel fuel," according to André-Jacques Auberton-Hervé, CEO of the electronic semiconductor manufacturer Soitec.[390]

Government officials usually justify energy subsidies by claiming that they help the poor, but the evidence shows that subsidies help the rich, not the poor. According to the IMF, the richest 20 percent of households in low- and middle-income countries net six times more in total fuel product subsidies than the poorest 20 percent of households.[391]

One billion Indians have access to a cell phone, but only 366 million have access to a toilet, according to the United Nations.[392] The country's leadership has to admit that it is not very good at providing infrastructure to its people even when doing so is rather simple. The toilet-vs-cell phone numbers prove that setting up a bit-based distributed infrastructure is easier than setting up an atom-based infrastructure consisting of water and sewerage pipelines, centralized processing plants, and command-and-control management.

What would it cost to provide solar electricity to 500 million Indians who currently have no access to the grid?

Assume that each Indian citizen gets about 100 solar Watts. A family of three would get 300 Watts, or about one solar panel; a family of five would get 500 Watts. India gets on average about five hours of sunshine every day. This means each family of five would get about 2.5 kWh per day, enough to charge a couple of cell phones and to run a computer, a television set, several LED light bulbs, a table fan, and a coffee pot.[393]

The 2.5 kWh per day figure is more than most grid-connected Indians consume. In 2005, 45 percent of Indians had no access to the grid at all; 33 percent had access to the grid but consumed less than 50 kWh per month (1.6 kWh per day); and 11 percent had access to the grid but consumed only 50 kWh to 100 kWh per month (1.6 kWh to 3.3 kWh per day). Eleven percent of Indians had access to the grid and consumed more than 100kWh per month (3.3 kWh per day).[394]

The cost of solar PV today is about $0.65 per Watt. That's just for the panel. Adding the costs of inverters, cables, installation, and so on, to the cost is about $2/Watt, which is the total installed cost per Watt for residential solar in Germany. Add a small battery for nighttime usage and you have a solar home system for about $3/W.

To provide solar electricity to 100 million people, India would need about $30 billion, less than what the government spends to subsidize diesel and other fossil fuels today. In other words, if the Indian government diverted the amount it spends to subsidize fossil fuels to home-based distributed solar generation, the country could bring electricity to 500 million people in just five years. The diesel and kerosene industries would be totally disrupted. There would be no need for new coal plants, hydroelectric plants, or transmission lines.

Solarizing the bottom 500 million in India would not only cost less than subsidizing  fossil fuels, it would end the unnecessary carnage caused by indoor air pollution. According to the World Health Organization, three to four-hundred thousand people in India die of indoor air pollution and carbon monoxide poisoning every year because of biomass burning and the use of chullahs.[395]

For the 1,411 people in tiny Tokelau or 500 million people in India, energy poverty can be ended quickly. Solar is already cheaper than diesel, kerosene, and many times more valuable than firewood. When governments wake up and stop subsidizing fossil fuels, the disruption of petroleum as an energy source will be inexpensive and swift.

India does not even need to subsidize solar for the energy disruption to happen. It just needs to stop subsidizing and protecting petroleum, coal, hydro, and nuclear. It also needs to change the regulatory apparatus that protects the incumbent command-and-control energy actors.

In 1991, to provide telecommunications for all, India decided to leapfrog the obsolete landline telephone model. Two decades later, nearly one billion Indians enjoy access to state-of-the-art telephones. Similarly, the disruption of energy in India will take place when the Indian government decides it is time to stop subsidizing and protecting its obsolete energy architecture in favor of allowing entrepreneurs to build a reliable, participatory, secure, and clean energy future.

# When Solar and the
# Electric Vehicle Convergence...

After the British Petroleum oil spill in 2010, I wondered about the combination of electric vehicles and solar. One of the myths about solar is that it takes up too much land. I thought I dispelled that myth in my book Solar Trillions, but apparently the fossil fuel industry is too good at misinforming the public. I'll do the numbers differently this time and hope that readers tweet the facts to overcome the fossil fuels propaganda machine.

The question before us is how much solar surface is needed to power all EVs in the United States for a year. After I make this calculation, I'll show how much drilling land is needed to power all internal combustion engine cars.

## How much solar surface area is needed to power all EVs?

Assume that every car and truck in the United States is electric. Assume furthermore that all power is generated by solar. How much surface area is needed to generate enough solar energy to power every car-mile in America?

Americans drive around three trillion miles (4.8 million Km) per year, according to the U.S. Department of Transportation.[396] How much solar energy is needed to generate enough power for EVs to drive those three trillion miles?

To answer that question, you need the number of miles per unit of energy (kWh) for an average electric vehicle. The Department of Energy compiles mileage figures for all electric vehicles in the American market.[397] The DOE measures how many energy units (kWh) it takes for a car to drive 100 miles.

The list includes Sports Utility Vehicles such as the Toyota RAV4 EV, large cars such as the Tesla Model S, and compacts such as the Ford Focus EV. For my calculations, I did a simple mileage average of all cars on the DOE's list.

The Nissan Leaf, the world's best-selling electric vehicle, gets 3.45 miles (5.5 Km) per kWh. Tesla's electric vehicles get better than 4 miles (6.4 Km) per kWh of energy stored in their batteries (see Chapter 5). These numbers are likely to improve as technology improves in the future. In fact, it is already possible to improve the Nissan Leaf efficiency up to 5 miles per kWh by changing driving assumptions.[398] For purposes of my calculations, I'm going to use Nissan's reported 3.45 miles (5.5 Km) per kWh.[399]

To calculate how much energy is needed to power all electric vehicles, divide the total number of miles driven (3 trillion) by the mileage (3.45 miles per kWh). A total of 869.6 billion kWh is needed to power all electric vehicles in the United States for one year.

The average sunlight-to-power efficiency for solar panels is about 16 percent. Sunlight-to-power efficiency measures, in percentage points, how much of the solar energy that hits the panel is converted to power. What the sunlight-to-power efficiency is depends on the technology used. At the low end of the market, thin film photovoltaic converts around 12 percent of solar energy to power, polycrystalline PV converts around 16 percent, and monocrystalline converts close to 20 percent. Concentrating solar power (CSP) technologies may double these numbers. Concentrating photovoltaics (CPV) has set records near 36 percent, while thermal CSP with combined heat and power (CHP) can reach an efficiency of 75 to 80 percent.[400]

The next step is calculating how much solar energy (insolation) that hits the panel every year. Solar Insolation is the actual solar energy to fall on a surface area expressed in $kWh/m^2/yr$. Solar insolation depends on the location of the solar plant. A solar plant in Barstow, California, may receive more than 2,700 $kWh/m^2/yr$; in Las Vegas or Tucson, Arizona, a solar plant may receive about 2,560 $kWh/m^2/yr$. My not-so-sunny hometown of San Francisco gets about 1,785 $kWh/m^2/yr$.

Assuming the solar plant is built in a desert somewhere in the southwestern U.S. and it gets 2,400 $kWh/m^2/yr$, about 874 square miles of land is needed to generate enough power for EVs to drive those three trillion miles. That's a square with 29.6-mile sides. In summary: About a thousand square miles of solar are needed to power every single electric car-mile driven every year in America. A solar plant the size of King Ranch in Texas with its 1,289 square miles could generate all of America's electric vehicle power with 40-percent extra electricity to spare.

Now let's compare one thousand square miles of solar with oil.

## Oil and gas land and water surface use

According to the U.S. House of Representatives, oil and gas companies leased 74,219 square miles (47.5 million acres) of land in the United States for the purpose of drilling oil. They leased another 68,750 square miles (44 million acres) for offshore drilling.[401] Altogether, the oil and gas industries lease about 143,000 square miles from the U.S. government — to meet just a third of America's transportation needs. Multiply 143,000 square miles by three and you get the total number of square miles the oil and gas industry needs to power every single vehicle-mile in America: about 400,000 square miles.

To power just one third of the gasoline car-miles in America, oil uses 143 times the surface area that solar would need to power all electric car-miles. Here's another way to look at it: The combination of solar and electric vehicles used land 400 times more efficiently for energy production than the combination of oil production and gasoline vehicle usage.

When a technological convergence (solar and EVs) is 400 times more resource-efficient than incumbent technologies (oil and internal combustion engine cars in this case), it's time to pay attention. The solar and EV technology conversion is bound to be disruptive, especially when you remember that, in this case, the resources are as valuable as land and water.

Of course, building a solar power plant that is 874 square miles is not feasible. Nor is it productive. The disruptive potential of solar technology lies not just in its low cost but in its distributed nature. It's better to generate most of that power close to the cars themselves from residential and commercial rooftops, malls, big-box stores, parking lots, landfills, and so on.

Wal-Mart is expected to cover 218 square miles in 2015.[402] Wal-Mart alone could power one fourth of all electric car-miles in the United States. All Wal-Mart would need to do is cover its rooftops with solar panels and its parking lots with solar canopies.

# Leaks, Spills, and Contamination

Needless to say, oil drilling leaks and spills damage more land and water than the numbers reveal. The 2010 BP Gulf Oil disaster damaged tens of thousands of square miles beyond its oil wells. As of June 2010, the U.S. National Oceanic and Atmospheric Administration (NOAA) Fisheries Services closed the 80,000 square miles around the wells to commercial fishing.

The BP Oil spill alone was eighty (80) times larger than the surface area that solar plants would need to power every electric vehicle car-mile in America. Meantime, no one has ever heard of a solar spill. Oil is not just dirty and expensive. Oil is a land and water hog. The convergence of electric vehicles and solar would be 400 times more land-efficient than oil and wouldn't threaten the kind of pollution that oil is famous for.

## Summary: The End of Oil

Oil is obsolete. The oil age will end by 2030. Electric vehicles, autonomous cars and solar energy will disrupt the oil industry. Most of the multi-trillion dollar investments in the oil industry will soon be stranded.

The oil industry will be disrupted and oil will become obsolete for many reasons. The data is overwhelming:
- Solar has improved its cost position relative to oil by 5,355 times since 1970.
- The convergence of solar and the electric vehicle is four-hundred times more land efficient than the combination of oil and the internal combustion engine.
- Solar is a distributed resource. It can power electric vehicles on-site without the need for expensive and inefficient pipelines, railways, cargo shipping facilities, refineries, storage facilities, and gas stations.
- Solar power is already cheaper than diesel for power generation.
- Solar power is already cheaper than kerosene for heat and lighting.
- Solar is already cheaper than petroleum for water desalination.
- Solar with 24/7 battery storage is already cheaper than diesel generation at small scale.
- Solar with 24/7 solar salt storage is already cheaper than petroleum generation at large scale.
- Electricity storage is going down so fast that, by 2020, solar PV with grid storage will be cheaper than petroleum anywhere at any scale.

The disruption of petroleum is already fast underway. The electric vehicle will disrupt the oil industry by 2030 — maybe before. The increase in oil demand over the next few years from energy-intensive growing economies like China and India will mask the fact that oil is on its way to quick obsolescence. Solar will disrupt oil as a power generation source and as an automotive power source. Finally, the autonomous vehicle will shrink the world's vehicle fleet and make it far more efficient. Whatever remains of the oil industry will be obliterated by 2030.

# Chapter 8: Natural Gas. A Bridge to Nowhere

*"We are shaping the world faster than we can change ourselves,*

*and we are applying to the present the habits of the past."*

*- Winston Churchill.*


*"You can fool all the people some of the time, and some of the people*

*all the time, but you cannot fool all the people all the time."*

*- Abraham Lincoln.*


*"In a time of universal deceit, telling the truth is revolutionary."*

*- George Orwell.*

On September 10, 2010, residents of the Crestmoor neighborhood in San Bruno, California, woke up to a natural gas pipeline explosion so loud and powerful it felt like an earthquake. Eyewitnesses described a "wall of fire 1,000 feet high." The U.S. Geological Survey later classified the explosion as a magnitude 1.1 earthquake.[403] The explosion killed eight people, created a crater 167 feet (51 m) long and 26 feet (7.9 m) wide, and flattened a whole neighborhood (see Figure 8.1).

The San Bruno explosion was a reminder of a truth we ignore at our peril: Gas pipelines leak. When that gas is methane (natural gas), the result can be catastrophic.

*Figure 8.1—Natural gas devastation in San Bruno. (Photo source: Brocken Inaglory)*[404]

About a century before the San Bruno explosion, in 1906, an earthquake and subsequent fire destroyed the then most important cultural, financial, and trading center on the west coast of the United States: San Francisco. About 25,000 buildings were destroyed, three thousand people died, and as many as 300,000 people (out of a total population of about 400,000) were left homeless.[405] While the magnitude 7.9 quake was devastating, it is estimated that 90 percent of the destruction in San Francisco was due to fires caused by ruptured gas mains.

## Is Natural Gas Clean?

In the first half of the 20th Century a new material was all the rage. At the 1939 New York World's Fair it was advertised as a "magic mineral" and extolled for its "service to humanity." It was used in dozens of products, from buttons to telephones to electrical panels.[406] Heart surgeons used it for threads and it was considered safe enough to purify food and to include in toothpaste. The name of this magic mineral was asbestos.

Natural gas is the new "magic" fossil fuel. The oil and gas industry has cleverly promoted natural gas as a clean energy source. The industry keeps reminding us that a natural gas power plant generates about half the greenhouse gases of a coal power plant (see Figure 8.2). What the industry doesn't say is that, when released unburned into the atmosphere, methane (the main component of natural gas) is 72 times worse than CO2 as a greenhouse gas (when measured over twenty years).[407]

*Figure 8.2—Advertising in Washington, DC: "Thanks Natural Gas For Less CO2" (Photo: Tony Seba)*

Natural gas leaks throughout the supply chain. It leaks when it is lifted from the ground, when it is stored, and when it is transported in hundreds of thousands of miles of pipelines. According to the U.S. Environmental Protection Agency, three trillion cubic feet of methane leak annually. That figure represents about 3.2 percent of global production.[408] This methane leakage is the global warming equivalent of half the coal plants in the United States.[409]

The gas line that ruptured and caused the San Bruno explosion in 2010 was a steel pipe with uneven wall thickness installed in 1956.[410] Federal investigators reported finding numerous defective welds. As growing volumes of natural gas increased the pressure on the pipelines, the pipelines were weakened until they blew up.

Oil and gas companies are racing to build new pipelines. More than four thousand miles of new pipelines were built in 2008 alone. However, most of the distribution of gas still takes place in older pipes. Over half of the 305,000 miles of natural gas pipelines in the United States was built in the 1950s and 1960s.[411] Over 12 percent was built during or before the 1940s, according to the U.S. Department of Transportation.[412] The San Bruno pipeline was built during this natural gas construction boom. The first federal pipeline safety regulations for gas pipelines were not put in place until 1968.

Natural gas volume in the United States is expected to increase by 50 percent over the next twenty years, according to the U.S. Department of Energy.[413] As volume and pressure increase on tens of thousands of miles of old rusty pipes that were laid out before the ink was dry on the first federal safety regulation, how much methane leakage can we expect?

There is surprisingly little data on pipeline leaks because pipeline owners are not required to report such data. The first-ever reported survey of pipeline leaks in the U.S. found 3,356 leaks along 785 miles of Boston roads.[414] That translates to 4.3 leaks per mile. Extrapolating this ratio, the United States may have 1.3 million natural gas leaks across 305,000 miles of natural gas pipelines.

The same research team that investigated Boston, led by Prof. Robert Jackson of Duke University, recently found 5,893 leaks in the Washington, DC natural gas pipeline system.[415] Some gas manholes had methane concentrations as high as 500 thousand parts per million — ten times higher than the threshold at which explosions occur. "The average density of leaks we mapped in the two cities were comparable, but the average methane concentrations are higher in Washington," said Nathan G. Phillips, a research team member and professor at Boston University's Department of Earth and Environment. Some of the leaks were comparable to the amount of natural gas used by two to seven homes; a dozen of the leaks pose a risk of explosion.

Methane, a greenhouse gas, is 72 times more potent (over twenty years) than $CO_2$.[416] Even a one-percent leakage rate would negate the fact that methane emits 50-percent less $CO_2$ than coal when burned in an efficient power plant. Even if you take the longer view and consider that methane is just 25 times worse than coal over a hundred years, a 3-percent leakage rate would negate its warming benefits versus coal.

On Halloween Day in 2013, more than three years after the San Bruno natural gas explosion tragedy, PG&E took out a page in the San Francisco Chronicle to proudly announce that it had installed ninety new automated valves and replaced 69 miles of its 6,750 miles of gas transmission pipelines.[417] It took the largest utility in the United States (by market capitalization) $2 billion and three years after a tragic explosion to replace a bit more than one percent of its gas pipeline. What will it take for PG&E to upgrade the 99 percent of the pipelines it didn't fix?

According to PG&E, it cost just under $29 million per mile to upgrade 69 miles of gas pipelines. Extrapolating these costs, rate payers would need to spend $193.7 billion to upgrade the 6,681 miles that PG&E didn't fix. At the end of 2012, PG&E had 4.4 million natural gas distribution customers.[418] Who would pay for these upgrades if they were to take place? Customers, of course. How much would it cost? It would cost each PG&E customer $44,000 for PG&E to fix its entire gas pipeline system. So much for "cheap gas."

The relentless advertising and lobbying campaign to convince the world that natural gas is clean has indeed worked. In fact, many municipalities, from San Jose, California to Seville, Spain, proudly reinforce the message that their "clean air bus" is powered by natural gas (see Figure 10.2).

*Figure 8.3—A "clean air" bus in San Jose, powered by natural gas (Photo: Tony Seba)*

By we know that natural gas is not clean. When I say that the future of energy is about clean, distributed energy, I want there to be no doubt: Natural gas isn't in the clean energy future.

## Is Natural Gas Cheap?

In March 1999 a barrel of oil cost $11.[419] As the 20th Century drew to a close, energy experts said we were entering an era of cheap and abundant oil. The Economist magazine predicted that a barrel of oil was headed to $5; Algeria's energy minister declared that oil might tumble to $2 or $3 a barrel.[420]

One barrel holds 42 gallons of oil. U.S. refineries produce 19 gallons of gasoline from each barrel of oil, according to the Energy Information Agency.[421] In 1999, the cost of gasoline was headed to 10 to 25 cents per gallon. Double that for profits before gas reaches the consumer and you were still looking at pump prices in the 20 to 50 cents per gallon (5.3–13.2 cents per liter) range.

Detroit was building the thirstiest gas-guzzlers in history. In a sign of the times, General Motors acquired Hummer, the then-ultimate gas-guzzling status symbol. It seemed that everyone in the automotive and energy business — even those who had access to the most detailed information and analyses money could buy —was confident in the idea of a cheap energy future.

History, of course, turned out differently. The market price for oil rose quickly as the new century drew to a close. It continued to rise and hit a high of $147 in July 2008 (see Figure 8.4).[422] In about eight years, the price of oil went up by fourteen times. The price rose to nearly 30 times The Economist's prediction and more than 50 to 70 times the Algerian energy minister's. Oil and gas markets are very volatile. Because they are based on resource commodities, prices can turn on a dime.

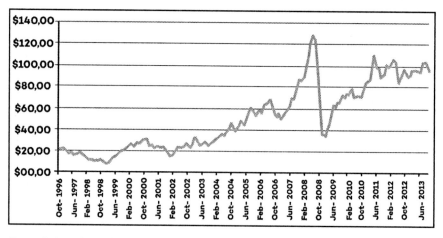

*Figure 8.4—U.S. crude oil first purchase price, 1996–2013. (Data source: EIA)*[423]

Hydraulic fracturing (fracking) has changed natural gas markets. Wholesale gas prices in the United States have dropped to twenty-year lows. The same industry experts who predicted a new millennium of cheap and abundant oil in 1999 now predict a new millennium of cheap and abundant natural gas.

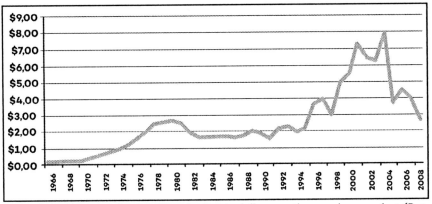

*Figure 8.5 – U.S. natural gas wellhead prices in dollars per thousand square feet. (Data source: EIA)*[424]

The wellhead price of natural gas has gone from $0.18 per thousand square feet in 1970 to $2.66 in 2012, according to the Energy Information Agency (see Figure 8.5). Despite the so-called fracking revolution, gas has gone up by a factor of 14.8 since 1970. Solar PV costs have dropped by a factor of 154 during the same period. Solar has improved its cost position by 2,275 times relative to gas since 1970.

The increased investment poured into fracking has cut the cost of gas from a high of $7.97 in 2008 to $2.66 in 2012. Gas has started displacing aging coal plants; it has gained market share in the power markets. The gas industry is partying like it's 1999 and promising a millennium of cheap energy. What is the likelihood that the cost of gas will stay this low for a few years or a decade?

Markets are not very good at predicting fossil fuel prices more than a few minutes into the future. Natural gas is particularly volatile (see Figure 8.6). Solar will have improved its cost by a factor of 400 between 1970 and 2020. Assuming gas stays at current levels until 2020, solar will still have improved its cost position relative to gas by a factor of 5,911 times relative to gas since 1970. Assuming gas reverts to its average price in the 2000–2012 period ($4.90), solar will have improved its cost position by 10,884 times relative to gas in the 1970–2020 period.

These are U.S. wellhead prices. Not every country partakes in the availability of shale gas. Only areas with the right geological conditions can take advantage of fracking technology. That is, the same countries that produce oil and gas may be able to produce more oil and gas. Everyone else will still have to import fracked oil and gas at expensive and volatile world prices.

Another point to consider is that gas is cheap mostly at the wholesale level. Distributing gas is expensive, even within a single country. In the U.S., the cost of exporting gas can be higher than the cost of extraction; exporting costs could wipe out any advantages from low domestic costs.

Figure 8.6 shows the prices of natural gas delivered to residential customers in the U.S. Notice a couple of things about this chart. First, prices are volatile. Volatility is increasing; the range of prices is getting wider within a given year. Within a two-year period, the price can double and then go down by half again. The second thing to notice is how prices fell in July 2008 after hitting a peak price above $20 per thousand cubic feet (the same time oil hit its own peak). However, despite low spot prices brought by the "fracking revolution" over the last few years, prices are still about twice what they were in the 1990s. Consumers are not getting the benefits of this so-called revolution.

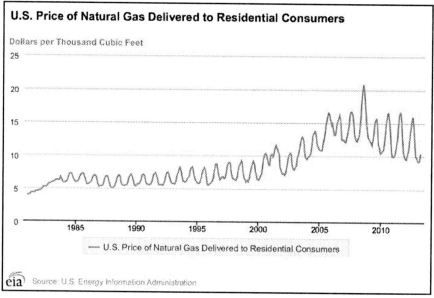

*Figure 8.6—U.S. residential prices of natural gas. (Source: EIA)[425]*

Contrast this with the prices of information technology or solar PV. Most of the benefits of the Internet have accrued to consumers, not producers. Again, when it comes to a sustained, multi-decade drop in prices, information economics and technology pricing beat resource economics.

## Solar vs. Natural Gas Prices

Unlike its liquid (oil) and solid (coal) fossil fuel siblings, natural gas is not easy to ship. This is why gas markets are regional. When natural gas hit its lowest cost level in the U.S. in 2012, import prices were five times higher in Europe and eight times higher in Japan, according to the International Energy Agency.[426]

In order to export natural gas by ship, the gas needs to be compressed or lique-fied. It must be shipped in special vessels and then decompressed or un-lique-fied (see Figure 8.7). Exporting and importing gas requires using compressed natural gas (CNG) or liquefied natural gas (LNG) facilities at both ports. The cost of these CNG and LNG facilities can be astronomical.

In Australia alone, oil and gas companies have poured A$ 200 billion (US$ 179 billion) into building liquefied natural gas plants.[427] Chevron disclosed that its Gorgon LNG plant would cost A$ 52 billion (US$ 46.6 billion.)[428]

Compare these costs to the cost of building solar PV power plants to meet the electricity demand in Australia. To make the comparison between providing Australia's electricity with natural gas or solar, I use these facts from the Aus-tralian energy industry:

- Australia's annual electricity generation: 241.6 TWh in 2009, according to the Australian Bureau of Resources and Energy Economics.[429]
- Solar insolation: 2,100 kWh/m$^2$/year.[430]
- Solar panel efficiency: 15.9%.[431]
- Installed cost per PV Watt: $1.62.[432]

The installed cost per PV Watt price may surprise Americans. It is far lower in Australia than the United States. On average for all of Australia, the total insta-lled cost of a 5kW residential system in July, 2013 was A$ 1.76 (US$1.62), accor-ding to SolarChoice. The cost in Perth was as low as A$1.38 (US$1.27/W).[433]

Using the average installed cost per PV, it would take an investment of US$186 billion for solar to generate all the electricity in Australia. Using the lower cost in Perth (US1.27/W), the investment would be US$146 billion.

Utility scale solar projects cost much less than residential solar, so the low-cost scenario would not be a stretch. Furthermore, the cost of solar is still dropping quickly. The final number would probably be lower than US$186.

Based on my calculations, the cost of providing all the electricity in Austra-lia with solar PV would be similar to (US$186 billion) or lower than (US$146 billion) the investment that the oil and gas industry is now pouring into a few liquefied natural gas (LNG) plants (US$179 billion).

Keep in mind the US$179 billion investment in question is just for liquefying natural gas. Liquefaction is just one step along the conventional energy value chain (see Figure 8.7). The US$179 billion figure doesn't include the cost of extraction, building pipelines to port, shipping, re-gasification facilities, and buil-ding more pipelines at the receiving end.

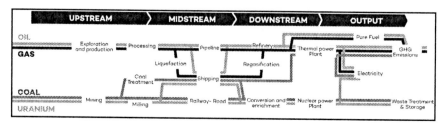

*Figure 8.7—Conventional (fossil and nuclear) energy value chain. (Source: IEA)[434]*

Each step along the value chain needs tens of billions or hundreds of billions of dollars in investments. And that's just to bring the natural gas from the ground to the power plant. Building the power plant, burning the gas to generate electricity, sending the gas over transmission lines and distribution lines, and sending the gas to homes adds further to the cost.

Natural gas follows a long and tortured path to get from underground to powering a home computer. If Rube Goldberg had been given an infinite amount of taxpayer money to create a power generator, he may well have conceived a fossil-fuel supply chain contraption.

Add it all up and natural gas is not just more expensive than solar, it's ridiculously more expensive, especially if you include the environmental costs that hydraulic fracturing extraction and pipeline leaks impose on air, water, and soil.

For instance, North Dakota fracking wells may produce as much as 27 tons of radioactive waste per day.[435] Shale rocks contain a high level of radium, a radioactive element used in cancer treatments. Radium-226 has a half-life of 1,601 years, which means that half of all that radium debris will still be emitting radiation sixteen centuries from now. Who pays for that contamination? You and I do. But not the oil and gas industry.

In Europe, where natural gas is imported (and expensive), gas cannot compete already in high-penetration solar and wind markets. Utilities in Europe have written down or written off investments in relatively new gas-fired power plants.[436] EON, Germany's largest utility, has been burning cash in two of its gas plants; these plants have been kept open only after getting special compensation deals with the grid operator.

When extracting fossil fuels using the hydraulic fracturing method, the oil and gas industry in the U.S. is exempt from many environmental laws, including:
• Clean Air Act
• Clean Water Act
• Safe Drinking Water Act
• National Environmental Policy Act

- Resource Conservation and Recovery Act
- Emergency Planning and Community Right-to-Know Act
- Comprehensive Environmental Response, Compensation, and Liability Act (Superfund)[437]

Essentially, when it comes pollution caused by extraction, the oil and gas industry is either above the law or writing the laws. The oil and gas industry has the right to pollute water, land, and air almost at will. When regulatory capture happens to the extent it has in the United States, disasters like the British Petroleum Gulf Oil spill are inevitable.

We have forgotten the words of Abraham Lincoln's Gettysburg Address: "The government of the people, by the people, for the people, shall not perish from the Earth." Instead we have "the government of the energy industry, for the energy industry, by the energy industry, shall have the right to destroy the earth."

These pollution costs are very real but are paid by the taxpayers, not by the industry that causes pollution. The industry doesn't have to disclose the names any of the hundreds of toxic chemicals that it pumps into the ground (and water) every time it "fracks" a well. We now know they pump radioactive radium out of the ground. I wonder if the industry pumps radioactive uranium or plutonium into those well holes. American citizens, even those who live in the areas being fracked, don't have the right to know.

## Water Conservation and the End of Natural Gas

It takes two to four million gallons of water to drill and fracture a single natural gas well using hydraulic fracturing methods.[438] Next time you read or hear about fracking, consider how much water fracking requires.

To date fracking has been performed more than one million times in the United States. In 2009, there were 493,000 active natural gas wells in the U.S.[439] It is estimated that, in 90 percent of these wells, fracking was used to get more gas flowing. In the state of Pennsylvania alone there are 150,000 abandoned oil and gas wells.[440]

Imagine how much water would be conserved if solar PV (or wind) generated the power that is now generated as a result of fracking. If we used solar PV (or wind) to generate the daily energy needs in the U.S., solar (or wind) would need about 11,000 $m^3$. That's about 2.9 million gallons of water.

To power the entire country with solar and wind would require the same amount of water as fracking a single natural gas well. Solar is literally a million times more water-efficient than gas. In terms of water usage, gas simply cannot compete with solar or wind.

When you take into account the pollution that fracking generates when the "fracked" water is dumped into streams, solar and wind look even better (see Figure 8.8). It is illegal to discharge this polluted water, but in the social media era in which we live it has been difficult for the industry to hide a practice that is embedded in its DNA. A search for "fracking illegal dumping of wastewater" gets more than 50,000 results on Google. Results include Bloomberg's "Exxon Charged with Illegally Dumping Waste in Pennsylvania"[441] and CBS's "Oil Company Caught Illegally Dumping Fracking Discharge in Central Valley."[442] The latter news story showed a video recorded by a farmer in Kern County, California. It showed an unconcerned drilling crew dumping illegal fracked wastewater in a river.

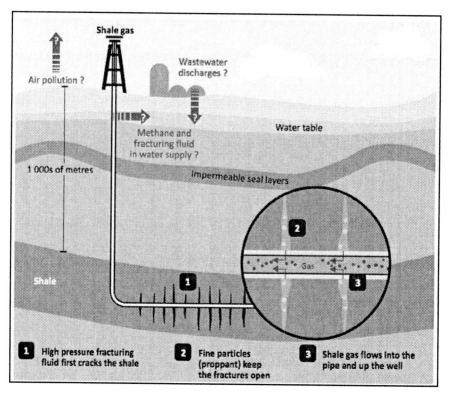

*Figure 8.8—Shale gas production technique and possible environmental hazards. (Source: IEA)*[443]

## Exponential Investments in Linear Returns

According to the International Energy Agency, oil demand will rise from 87.4 mb/d (million barrels per day) in 2011 to 99.7 mb/d in 2035.[444] Most of the new oil will be "unconventional" oil extracted by fracking, oil sands, and other unconventional technologies. That's a 0.5-percent compounded annual growth rate (CAGR).

In a report auguring a "golden age" of gas, the International Energy Agency forecasts a rise in natural gas production from 3,721 bcm (billion cubic meters) in 2012 to 5,112 bcm in 2035 for a total growth of 37.4 percent.[445] That's a 1.37-percent compounded annual growth rate (CAGR). Between 2010 and 2035, the oil and gas industry will spend US$15 trillion in upstream activities (exploration and drilling) to meet the increasing world demand for oil and gas.[446] Between 2010 and 2035, during what it calls a "golden age," the oil and gas industry will invest more than the total gross domestic product of the United States to deliver a whopping 0.5- to 1.4-percent growth per year.

Forgive me if I'm not impressed. The oil and gas industry promises a "golden age," but the "golden age" requires massive, multi-trillion dollar investments. It requires untold amounts of sand, water, and who-knows-what chemicals. It requires society to bear frightening and burdensome environmental costs. And the end result is at best 1.37-percent annual growth.

Would any business executive invest $15 trillion of her company's money to achieve one percent growth? A look at the International Energy Agency report explains why the answer may be "yes" if the executive worked in the oil and gas industry. The report shows that the fossil fuel industry received annual subsidies of $523 billion in 2011.[447] Over the period 2010–2035, this would amount to $13.5 trillion, about 90 percent of the $15 trillion that the oil and gas industry will need to invest in that period.

The oil executive's spreadsheets work after all: Taxpayers finance 90 percent of the capital investment for drilling oil and gas; the drilling takes place mainly on public lands using publicly owned water; the drillers are exempt from any damage to the air, water, and soil. Taxpayers bear the risks and costs and the industry takes trillions in profits. This formula generates unbelievable returns for the oil and gas industry.

Natural gas, the new "magic fossil fuel," is a bridge to nowhere. It's a destructive, resource-inefficient, financially unviable source of energy. Government protection, exemptions from the rule of law, plus trillions of dollars in taxpayer subsidies keep natural gas flowing.

# Chapter 9:
# The End of Biofuels

*"The future isn't what it used to be"*

- Paul Valéry

*"There is nothing more frightful than ignorance in action"*

- Johann Wolfgang von Goethe

*"In God we trust. All others bring data."*

- W. Edwards Deming

On June 18, 2011, Honeywell announced the first transatlantic flight powered by biofuels. The company flew a Gulfstream G450 from Morristown, New Jersey to Paris, France. The Gulfstream G650 was powered by a "50/50 blend of Honeywell Green Jet Fuel and petroleum-based jet fuel."[448] Honeywell said about the fuel it used for the flight: "Honeywell has produced more than 700,000 gallons of Honeywell Green Jet Fuel from sustainable, inedible sources such as camelina, jatropha and algae for use in commercial and military testing."

Biofuels (especially ethanol) have been used interchangeably with the words renewable and sustainable for at least three decades. During the 2012 U.S. Presidential debates, both candidates, Mitt Romney and Barack Obama, used similar language to express support for "renewable energy." "I believe very much in our renewable capabilities — ethanol, wind [and] solar will be an important part of our energy mix," said Romney, mirroring Obama's position on the issue.[449]

What does "renewable" mean? According to the U.S. Energy Information Agency (EIA), "renewable energy sources regenerate and can be sustained indefinitely."[450]

We know the sun will shine indefinitely (at least one billion more years!).[451] We know the wind will blow indefinitely.

Can agricultural biofuels "regenerate" and be "sustained" indefinitely? Is that Gulfstream biofuels flight to Paris "sustainable" or "renewable"? Can we grow biofuels indefinitely? The evidence gives a definite answer to these questions: NO.

This chapter looks at the evidence. Are biofuels "renewable"? Do they have a role in the clean disruption?

## Squandering Water Resources with Biofuels

Water is energy and energy is water. Energy is needed to pump, clean, and transport water. Water is used to mine, wash, and generate energy. With agricultural biofuels, water is used to "grow" energy.

The entire thermal energy industry relies on massive amounts of water. About 15 percent of the world's freshwater withdrawal goes for energy, according to the World Bank.[452] In a world where nearly half the population will live in areas of "high water stress," demands on the fresh water supply by the thermal energy industry will be especially taxing. It takes 13,676 gallons of water to produce a single gallon of biodiesel from soybeans, according to *WaterFootprint*.[453]

It would take 820,560 gallons of water to produce enough soybean biodiesel to fill up a 60-gallon biodiesel bus. To put this in context, an Olympic-size swimming pool holds approximately 660,000 gallons of water.[454] To fill its tank with soybean biofuel, a "clean and renewable biodiesel" bus like the one in Figure 9.1 would use enough water to fill an Olympic swimming pool.

*Figure 9.1—"This bus runs on biodiesel fuel," but the fuel isn't clean or renewable. (Photo: Tony Seba)*

How many gallons of water are needed to fill up an SUV with corn ethanol? I put this question to Prof. David Pimentel of Cornell University. He has studied biofuels for more than three decades and a foremost authority on water and biofuels.[455] "It takes 1,700 gallons of water to produce one gallon of ethanol," Prof. Pimentel answered. "Assuming a 30-gallon gasoline tank in a SUV, we calculate 51,000 gallons of water for the corn ethanol fill up."

According to the United States Geological Survey, the per-capita residential consumption of water in the U.S. in 2005 was about 99 gallons per day, down slightly from 101 gallons per day in 1995. Daily water usage ranged from 51 gallons per day in Maine to 189 gallons per day in Nevada.[456]

On the basis of these numbers, each time someone fills up an SUV with corn ethanol, he or she uses as much water as the average American residential user consumes in 16 months!

All forms of energy generation use water in the production process. However, the amount of water that is used varies by orders of magnitude. IBM's "Carbon Disclosure Project" gives an indication of the differences (see Table 9.1). To generate 1 MWh of energy (about what an American home uses each month):
   • Solar PV and wind use negligible amounts of water: 0.1 liters or less than half a cup.

- Natural gas uses ten thousand times more water than solar to generate the same 1 MWh of energy.
- Nuclear and coal use about twice as much as gas.
- Oil uses twice what coal uses (and forty thousand times the water solar uses).
- Hydroelectric power uses 680,000 times more water than solar.
- Biofuels use an astonishing 1.78 million times more water than solar to generate the same amount of energy.

| | Total Water consumed per megawatt-hour (m3/MWh) | Water required for US daily energy production (millions of m3) |
|---|---|---|
| Solar | 0.0001 | 0.011 |
| Wind | 0.0001 | 0.011 |
| Gas | 1.0000 | 11.000 |
| Coal | 2.0000 | 22.000 |
| Nuclear | 2.5000 | 27.500 |
| Oil | 4.0000 | 44.000 |
| Hydropower | 68.0000 | 748.000 |
| Biofuel (1st-gen) | 178.0000 | 1198.000 |

Table 9.1—Total water consumed per MWh for different sources of energy. (Source: IBM, "Carbon Disclosure Project Report")[457]

Here's another way to look at it: It would take 1.2 billion cubic meters ($m^3$) of water for biofuels to produce all the *daily* energy needs in the U.S. China and India together consume about 2.4 billion $m^3$ of water *per year*.[458] That is, all the water for drinking, agricultural irrigation, power plants, and factories that 2.5 billion people in China and India need per year is roughly equal to the water biofuels require to produce two days' worth of energy in the U.S.

The whole world consumes about 9 billion $m^3$ of water per year.[459] To produce about one week of America's energy needs, biofuels would suck up all the freshwater that our whole planet consumes in a year. It would take just one week of American biofuel energy production to turn earth into a planetary Sahara desert.

Compare that with solar PV or wind. To generate the daily energy needs in the U.S., solar (or wind) would need about 11,000 $m^3$. That's about 2.9 million gallons — less than five swimming pools.

Fracking a single well takes more water than that. It takes up to 4 million gallons of water to drill and fracture a single natural gas well using hydraulic fracturing methods.[460]

It just doesn't make any sense to use a resource as valuable as water the way most of our energy sources do. In an era of record temperatures, a rising population, and growing water needs for drinking and food production, biofuels cannot be part of a viable energy strategy. Agricultural biofuels are a true environmental catastrophe in the making.

## How Much Water Is Needed for Biofuels?

But isn't water renewable? Isn't there a natural water cycle that brings water back? If that were the case, we could consume water indefinitely and water would indeed be renewable.

To answer the question of whether water is renewable, let me introduce you to the most important source of water most Americans have never heard of: the Ogallala Aquifer.

### America's Aquifer

The Ogallala is one of the largest and most plentiful underground freshwater "oceans" in the world. It runs the length of the United States from South Dakota to Texas, a total of 174,000 square miles (450,000 km²). The Ogallala Aquifer is about 20 percent larger than all of Germany. It is what makes the U.S. Midwest an agriculture capital of the world: 30 percent of all the irrigation water in the United States is pumped from the Ogallala Aquifer.[461] It also provides drinking water for 82 percent of the people who live within its boundaries.

The aquifer recharges at the rate of 0.024 inches (0.61 mm) to 6 inches (150 mm) per year, but is being pumped out at industrial scale rates. In some areas, the water table has dropped five feet (1.5 meters) every year. Altogether, the Ogallala is probably being depleted at a rate of about 9.7 million acre-feet (12 km³ or 420,000 million cubic feet) annually, which is equivalent to eighteen Colorado Rivers pouring out to the sea every year.[462]

Is Ogallala water "renewable"? Not at the rate it is being depleted. Can we keep pumping water from the Ogallala indefinitely? No. The Ogallala Aquifer may dry up in our lifetimes. Some estimates give it twenty-five years.[463] By that time, the agricultural biofuels industry may have succeeded in turning the breadbasket of the world into a massive desert.

## Atlanta in Your Fuel Tank

At the start of this chapter, I described Honeywell's flight across the Atlantic in a Gulfstream G450 and how this flight was powered by biofuels. How renewable or sustainable was Honeywell's flight from New Jersey to Paris? Let's do the numbers.

A Gulfstream G450 carries up to 4,402 gallons of fuel.[464] The Gulfstream G450 was fueled by a mixture with 50 percent biofuels, which puts the biofuels it carried at about 2,201 gallons. Honeywell said its Green Jet Fuel was composed of jatropha, camelina, and algae. For the purposes of this exercise, we will examine jatropha. According to WaterFootprint, 19,924 gallons of water are needed to produce one gallon of biodiesel from jatropha. Based on that number, 43.8 million gallons (166 million liters) of water were needed to produce enough biofuel to power half of the Gulfstream G450's flight from New Jersey to Paris.

To put that number in perspective, producing 2,201 gallons of biodiesel from jatropha requires an amount of water equal to the daily water consumption of 442,956 Americans. In other words, producing enough biofuel from jatropha for a transatlantic flight requires the amount of water needed daily by the people of Atlanta (420,003).[465]

A flight from New Jersey to Paris in a small Cessna airplane that carries 14 to 19 people half-filled with biofuels requires an amount of water equivalent to the daily residential water consumption of a city the size of Atlanta. This is an insane amount of water.

I showed these figures to Prof. Arjen Y. Hoekstra of the University of Twente (Netherlands). The professor is the creator of Water Footprint, and one of the world's foremost authorities on water management. "Your numbers are correct," he wrote in an email. "They show that we need to carefully look at how we use limited resources like freshwater. Those who want to generate sustainable energy should look at solar and wind. If you want to do biofuels then you should start with bio waste, not agriculture."

On January 16, 2012, Lufthansa proudly announced the completion of the first transatlantic biofuel flight from Europe to the United States.[466] According to Boeing, a 747-400 has a fuel tank capable of carrying 57,285 gallons (216,840 liters.)[467] The biofuel Lufthansa used was made from a combination of camelina oil from the U.S., jatropha oil from Brazil, and "some animal fat" from Finland. For the sake of discussion, assume that 100 percent of the fuel tank was filled with jatropha biofuel.

It takes 19,924 gallons of water to produce one gallon of biodiesel from jatropha. It would take 1.14 billion gallons (4.3 billion liters) of water to produce the jatropha biofuel to power the flight from Frankfurt to New York.

In Germany, the water consumption per person is 122 liters per day.[468] So that one-way 747 Lufthansa flight from Frankfurt to New York used the same amount of water as 35.4 million Germans consume daily. Lufthansa would certainly never use biofuels if it had to grow them in Germany. Does anyone at Lufthansa really think biofuels are sustainable?

Scientific evidence says agricultural biofuels are not sustainable or renewable. But politics says they are.

## What about Next-Generation Biofuels?

Often when I show the above calculations to "biofuel believers" I am greeted with disbelief. That's normal. A few seconds later a question inevitably comes back: "Yes, but how about next-generation biofuels?" "Next-generation biofuels" usually means "cellulosic biofuels,".

Conceptually, the main difference between "first-gen" and "next-gen" biofuels is that the former uses the "edible" sugars or oils part of the plant, whereas the latter are made by breaking down the cellulose components of the rest of the plant, such as leaves, stems, and other fibrous parts of the plant.[469] I say "edible" in quotes because certain "popular" biofuels such as jatropha are not only not edible, they're actually poisonous.[470]

"Next-generation biofuels" seems to have borrowed its strategy from the coal lobby's "clean coal" campaign. Both of these slogans are meant to keep the financial subsidies and political protection flowing, the faithful fighting, the unaware hoping, and the scientific evidence at bay.

Despite massive taxpayer subsidies and consumption quotas imposed by governments to create a market for biofuels, companies are exiting the industry and investors are exiting along with them. Investors don't believe the biofuels industry will go anywhere anytime soon.

After investing hundreds of millions of dollars in two companies to try to make biofuels work, Al Shaw, CEO of Calysta Energy, gave up. His company switched to using natural gas as a source fuel. "Biomass doesn't cut it," Shaw said. "Carbohydrates are not a substitute for oil. I was wrong in that, and I admit it."[471]

Shaw's previous company, Codexis, invested $400 million from Shell Oil to develop ethanol from cellulosic, a "next-generation" biomass. Shell has announced it will not invest in biofuels research anymore. "That [cellulosic biomass] will never replace oil because the economics don't work," Al Shaw said. "You can't take carbohydrates and convert them into hydrocarbons economically."[472] BP also cancelled its biofuels plans.[473]

Wall Street investors don't expect much progress with next-generation biofuels either. Global investments in the biofuel industry dropped 99 percent, from a quarterly high of $7.6 billion in the fourth quarter of 2007 to just $57 million in the first quarter of 2013, according to Bloomberg New Energy Finance.[474]

As the evidence against biofuels mounts, Calysta Energy and other companies who were originally in the agricultural biofuels business have given up on agricultural biomass. They have switched to using natural gas as a feedstock instead of agricultural biomass.

The Rube Goldberg-like value chain of natural gas has now met the Rube Goldberg-like value chain of biofuels. The result is an intricately complex mesh of subsidies, government protection, and quotas that is hard to disentangle.

The ethanol market is a government-created support mechanism for the agricultural biofuels lobby. Ethanol prices are artificially inflated to reflect this support. Did politicians really intend the ethanol market to be another channel for natural gas?

## Why Solar Is More Efficient than Biofuel

Green plants — the type that grow in the garden — are solar plants. They convert solar energy into biomass (into wood, fruit, leaves, roots, and so on) with an efficiency of less than 0.3 percent. Furthermore, to convert this tiny percentage of solar energy into biomass, plants need massive amounts of help from other valuable resources, including water, land, and fertilizer.[475] Sugarcane, one of the biofuel industry's "success stories," converts just 0.38 percent of sunlight into biomass. To convert sugarcane into ethanol, it has to be planted, tended, harvested, transported to a refinery (see Figure 9.2), and treated with energy and more water. In the end, the solar-to-ethanol conversion ratio for sugarcane is just 0.13 percent, according to Scientific American.[476]

*Figure 9.2— Biofuels refinery. (Photo Source: Iowa Energy Center)*[477]

Compare that with the average solar panel's conversion rate of 16 percent. The average solar panel is 123 times more efficient in converting sunshine into usable energy. Furthermore, solar panels don't need fertilizers, water, pesticides, or extra energy to convert sunshine into electricity.

Concentrating photovoltaic (CPV) can turn about 40 percent of the sunshine into electricity, which makes it more than three hundred times more efficient than sugarcane ethanol biofuel. Other solar technologies, such as concentrating solar with combined heat and power (CHP), can convert more than 72 percent of the sunshine into usable energy.[478] Solar CHP can convert sunshine 550 times more efficiently than sugarcane can.

Solar is anywhere from 123 to 550 times more efficient than biofuels. Take any land in the world and PV technology will be at least a hundred times more efficient in turning sunshine into energy. PV will do it without using valuable land or water, let alone fertilizers or toxic pesticides that contaminate and create persistent runoffs.

Will there ever be a real market for agricultural biofuels? "The whole market for biofuels is 100-percent political," said Jeff Passmore, an executive VP of Iogen, a biofuels processor.[479]

Many government subsidy programs continue decades past the expiration dates of the products they help prop up. The telegraph, for example, has been obsolete for decades. It lives on only in museums and Western movies. Yet the

Indian government stopped running its telegraph service in 2013.[480] By that time there were 867 million cell phone subscribers in India.

Agricultural biofuels are a well-meaning experiment gone awry. They are already obsolete. The only thing that's renewable about the agricultural bio-fuels industry is the special-interest lobby groups that represents it in places like Washington, Brazilia, and Brussels. As the diamonds commercial would say: Lobbies are forever.

## The Hunger Games: Final Battle between Obsolete Biofuels and Oil

I mentioned in Chapter 8 that oil will be obsolete within two decades. Two disruption waves (the electric vehicle and the autonomous car) will shrink the demand for oil. The solar disruption wave will shrink the diesel power market to a shadow of its former self.

As demand for oil plummets so will the price of oil. Investments in expensive and inefficient production endeavors like offshore drilling and the Canadian Oil Sands will stop because they will not be financially viable. Only fields whose production costs are really low will remain in the market.

Biofuels will be even more uncompetitive; taxpayer subsidies will explode. Despite all the "free" water, production subsidies and consumption quotas, biofuels haven't been able to compete with oil when its price is at the $100/bbl level.

Assuming there's any freshwater to make biofuels in 2030, will the biofuels industry be able to compete at a price level 70- to 80-percent below the current market benchmark? Governments will commit taxpayer money to pay for the rising gap between the high production costs of biofuels and the decreasing market price of its competition: petroleum.

Oil companies don't take biofuels seriously as competitors anymore. They have decades of evidence to show that biofuels cannot compete with oil. However, as the oil industry is disrupted and shrinks in the future, the oil lobby will have to fight for every piece of the market it can get its hands on. Inevitably, the oil lobby will have to pick a fight with the agricultural lobby.

Who will win a serious head-on battle in 2030 between a shrinking, wounded oil lobby and a still powerful agricultural lobby? Two obsolete sources of energy will battle it out in the halls of political power.

The oil lobby will probably battle biofuels on environmental grounds. They will likely say that, when you take into account the whole production lifecycle (including fertilizers, energy, transportation, manufacturing, and so on):

- Oil generates less CO2 and other greenhouse gases than biofuels.
- Oil uses orders of magnitude less water than biofuels.
- Oil doesn't use prime agricultural land for energy production.
- Oil doesn't need as many taxpayer subsidies.

Two obsolete forms of liquid transportation energy that lost in the marketplace will fight each other in the alternative universe of politics. It will be like watching the jukebox and the 8-track tape industry lobby the government for a piece of the taxpayer pocketbook. The oil lobby will take the high environmental ground against agricultural biofuels. May we live in interesting times!

# Chapter 10:
# The End of Coal

*"It is not the lack of inventive ideas that set the boundaries for economic development but rather powerful social and economic interests promoting the technological status quo."*

*- Joseph Schumpeter.*

*"I am become Death, the destroyer of worlds."*

*- Robert Oppenheimer.*

*"Countries choose to live in the energy dark ages not because the sun fails to shine but because they refuse to see it."*

*- Tony Seba (paraphrasing Michener).*

On June 26, 2013, the World Bank announced it would stop financing new coal-fired power plants.[481] In a paper published on its website, the World Bank said it would finance coal only in the "rare circumstance" in which there was "no feasible alternative to coal and a lack of financing for coal."[482] The following month, the European Investment Bank (EIB) announced that it, too, would stop funding new and refurbished coal-fired plants. The EIB, which invests in the 28-country European Union bloc, said it would only invest if coal plants met certain emissions standards.[483]

U.S. President Barack Obama has called on American multilateral financing organizations to "end public financing of new coal plants overseas unless they deploy carbon capture technologies."[484]

These moves by the World Bank and European Investment Bank won't end coal's reign as the King of Power (electric and political.) The coal industry has been lobbying political institutions for three-hundred years. It knows how to manipulate governments and energy agencies. The coal industry has perfected the art of "regulatory capture."

Over the last century, the World Bank, European Investment Bank, Inter-American Development Bank, Ex-Im Bank, and similar financial organizations fell over each other to be first in line when it came to financing coal-fired plants. The World Bank, for instance, funded $6.26 billion worth of coal projects over the past five years alone.[485]

Because financial institutions were so eager to fund coal, the cost of capital to coal power plants has historically been artificially low. Since taxpayers fund these multilateral finance organizations, this is yet another way in which taxpayers have been subsidizing the coal industry around the world – by keeping the cost of capital artificially low.  This is about to change. Now that the World Bank, EIB, and similar organizations are reluctant to fund coal, the cost of capital for coal-fired plants will rise. Soon coal-fired plant producers will have to make more trips to private lending and investment markets to finance their operations.

Without generous government loans and loan guarantees as a backstop, Wall Street's lending standards will be more stringent. As a consequence, the cost of capital for coal power will go up. Finance 101 tells you that, as the cost of capital goes up, two things will happen in the coal industry:
   • Fewer coal plants will be built because they will not be financially viable. Many projects whose net present value (NPV) was positive due to artificially low capital costs will have a negative NPV and will be turned down.

- Coal projects that do get built will produce more expensive electricity. The electricity will be more expensive because the plants will have to pay a higher interest rate on their bank loans.

## Coal – a Risky Propostion

On August 21, 2013, less than two months after the World Bank announced it would stop financing new coal-fired power plants, something happened for the first time in coal history: The U.S. Bureau of Land Management (BLM) held a coal lease in Wyoming and no one bid.[486]

Traditionally, whenever the BLM offers land for coal mining, there is only one bidder, the company that originally applied for the land. In this case, Cloud Peak Energy applied to lease that land seven years before the BLM auction and was expected to be the sole bidder. "We were unable to construct an economic bid for this tract at this time," said Colin Marshall, the CEO of Cloud Peak Energy.

Was it possible this company could not find a market for 149 million tons of coal next to its Cordero Rojo mine? Or did Cloud Peak Energy not make a bid because the cost of capital to a once-invincible industry was rising?

Coal giant Walter Energy had to pull a credit refinancing of $1.55 billion[487] and saw its stock price drop by about half in 2013, according to Morningstar.[488]

At the end of 2013, the total combined market valuation of the select 32 coal companies in the Stowe Global Coal Index was $132 billion.[489] Facebook's valuation alone was $136 billion.[490] Google's valuation was $373 billion, nearly three times the total combined value of the 32 top coal companies in the Stowe Global Index.[491]

Wall Street used to consider financing coal projects a low-risk strategy, but reality has changed dramatically and risk perception is catching up with reality. Accordingly, the cost of capital for coal has already gone up. Walter Energy's inability to raise inexpensive debt (or any debt at all) tells us that Wall Street has already raised the cost of debt for coal projects. The low market valuations of coal companies signal that Wall Street doesn't foresee any growth any time soon.

Coal is becoming a risky investment and its cost of capital is going up accordingly.

Coal companies themselves are aware of those risks. In the second quarter of 2013, Cloud Peak Energy made more money from financial derivatives than selling coal.[492] The leading coal company earned more money by betting against the price of coal than it did by actually digging coal out of the ground. Companies betting against themselves could be seen as a red flag even if it's a generally accepted financial practice.

The coal industry is down. But is it out? Is Wall Street pointing to a long downward trend or is this a blip in the radar?

## Coal – a Death Foretold

The death of the coal industry in advanced industrialized economies has taken some by surprise, but the industry has been dying for decades. The coal power plant building boom started in the mid-1950s and peaked in the early 1980s, according to the U.S. Energy Information Agency (see Figure 10.1). In the 1970s, utilities started turning to the promised land of nuclear plants, which had a peak-commissioning era in the 1980s (see Chapter 8).

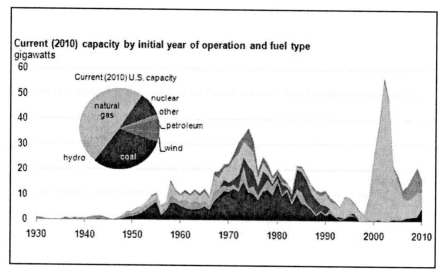

Figure 10.1— Generation capacity in the U.S. by initial year of operation and fuel type. (Source: Energy Information Agency)[493]

From 1990 to 2012, most new generation capacity in the U.S. came from natural gas power plants.. Figure 10.1 shows that the burst in the construction of new natural gas plants took place before the current "shale gas revolution." The reasons for this burst of construction were threefold.

The first reason for the revival of the gas market was the Natural Gas Utilization Act of 1987. This Act repealed sections of the Powerplant and Industrial Fuel

Use Act of 1978, which restricted the use of natural gas by electric utilities. The Industrial Fuel Use Act of 1978 was passed as a result of the first oil crisis of 1973. When this oil crisis hit world markets in 1973, 16.9 percent of electricity in the U.S. was generated using petroleum and 18.3 percent was generated in gas-fired power plants.[494] Concerns about tightening oil and gas supplies led the U.S. Congress to enact the Powerplant and Industrial Fuel Use Act of 1978. It banned the construction of new power plants that used oil or gas as the primary fuel. The Act also restricted the use of oil and gas for industrial use and encouraged the development of nuclear and coal-fired plants.[495]

By 1987, oil represented just 4.6 percent of U.S. electricity generation (down from 16.9 percent in 1973); gas had dropped to just about 10.6 percent (down from 18.3 percent in 1973). The Natural Gas Utilization Act of 1987 allowed industry and electric utilities to use natural gas again.

The second reason for the revival of the natural gas market had to do with two laws passed by the U.S Congress that helped to deregulate electricity markets. The Public Utility Regulatory Policies Act of 1978 (PURPA) broke up the vertical integration of utilities by allowing independent power producers to enter the electricity generation market. The Energy Policy Act of 1992 (EPACT) created the framework for a competitive wholesale electricity market.[496]

EPACT mandated the Federal Energy Regulatory Commission (FERC) to open up national electricity transmission systems and eliminate barriers that prevented non-rate-based power plants from entering these markets. Most of these new power plants were expected to be gas-fired.

The third reason for the revival of the natural gas market was technological. Gas turbine technology had steadily improved since the post-War War II era. The introduction of combined-cycle power plants essentially doubled the conversion efficiency of gas-fired power plants between 1970 and 1990 (see Figure 10.2). By the time EPACT was enacted in 1992, new combined-cycle gas-fired turbine (CCGT) power plants had thermal conversion efficiencies above 50 percent. Thermal conversion efficiencies have since gone up steadily to about 60 percent. Thermal conversion efficiency refers to the percent of the heat energy created by combustion that is actually converted to mechanical power and then electricity.

By contrast to CCGT power plans, a typical coal-fired plant has a thermal efficiency of about 33 percent. This means that two-thirds of the heat energy created in the combustion of coal is wasted.

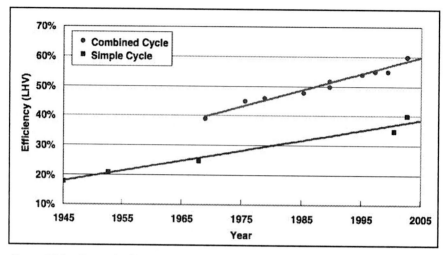

Figure 10.2—General efficiency increases over time for simple and combined-cycle gas turbines. (Source: Massachusetts Institute of Technology)[497]

For the three reasons outline here — the Natural Gas Utilization Act of 1987, which allowed industry and electric utilities to use natural gas again; the deregulation of utility markets; and improvements in gas turbine technology — a dramatic shift in power markets occurred. This shift benefited natural gas (and later wind and solar).

Most coal-fired plants in the U.S. are near or past retirement age. A typical coal-fired power plant lasts forty years. More than 540 GW (51 percent) of total generating capacity in the U.S. is older than thirty years, according to the Energy Information Agency (see Figure 10.3). More than 74 percent of all coal-fired capacity was older than thirty years. All this generation will have to be replaced over the next ten to twenty years. In fact, much of this generation is way past retirement age (about forty years) and is on maintenance life support.

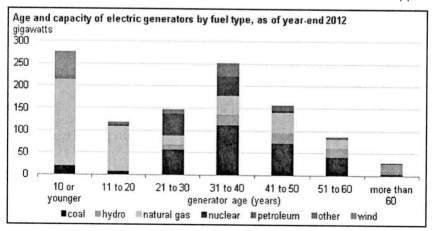

*Figure 10.3 – Age and generation capacity of existing electric generators in the U.S. by fuel type. (Source: Energy Information Agency)*[498]

What will replace this ageing coal and nuclear plant fleet? The evidence in the U.S. strongly suggests that coal and nuclear are on their way out. Most new capacity added over the two decades prior to 2012 has been natural gas and wind (see Figure 10.3). How do U.S. utilities plan to replace ageing power plants? NV Energy provides a case in point.

In April 2013, NV Energy announced its generating capacity investment strategy to 2025. There are two parts to the strategy, which it calls NVision:[499]
- Shutter its four coal-fired power plants. The plan calls for closing its three plants in Moapa in 2014 and its fourth plant in 2017, at which point NV Energy will have no coal in southern Nevada.
- Instead of coal, the utility will invest in natural gas and clean energy in a 60–40 split.

NV Energy had previously signed a 25-year agreement with SolarReserve to build a 110MW baseload (24/7) solar power plant in Crescent Dunes, Nevada. This solar power tower plant has ten hours of energy storage, so it can produce solar electricity on demand whenever customers need it.

Baseload, on-demand solar power has already proven itself in markets such as Spain, but it will be new in Nevada when the Crescent Dunes plant opens in early 2014. NV Energy's contract calls for purchasing peak solar power for 13.5 ¢/kWh, many times cheaper than natural gas peaking plants. Speaking to my class at Stanford University, Kevin Smith, CEO of SolarReserve, said that his company is looking to cut costs by half over the next five years.

Soon after NV Energy announced its new 12-year plan, the company was acquired by MidAmerican Holdings.[500] MidAmerican Holdings, a subsidiary of Warren Buffett's Berkshire Hathaway, had previously invested $2–$2.5 billion to acquire a solar development project that will be the world's largest solar power plant (579 MW) when it opens in 2015.[501]

Except for plants that are already under construction, "there are no new coal plants on the horizon in the United States," according to a U.S. Congressional Research Services Report.[502] The industry may blame new regulations, but the fact is that coal plants have accounted for less than 10 percent of new power capacity in the U.S. since the early 1990s.

Coal is dying not because it's dirty, but because it is uncompetitive. Coal in the U.S. is already obsolete. Coal power is quickly losing market share in the U.S to natural gas, solar, and wind.

But coal is not dead yet. Like a five-headed pyro-hydra, coal is rearing its heads elsewhere on the planet.

# Regulatory Capture: How Governments Protect the Coal Industry

Regulatory capture happens when a regulatory agency that is supposed to regulate an industry on behalf of the people instead regulates the people on behalf of the industry. In other words, regulatory capture is when regulators game the system to the benefit of the companies that they are supposed to regulate. Regulatory capture can permit companies to generate massive pollution with the full knowledge that the government will protect them and the taxpayers will pay for the clean-up costs. Regulatory capture is endemic in the conventional energy world.

Coal is dying in the U.S. as utilities switch to natural gas, wind, and solar, but coal generates between 40 and 50 percent of electricity worldwide and may well grow that percentage in the coming years. China alone consumed 46 percent of global coal in 2010.[503] Between India and China, about 1 Terawatt of new coal plants, the equivalent of all electricity generated from all sources in the United States, are waiting in the pipeline to be developed over the next few years.

China is the world's second largest subsidizer of energy on a post-tax basis, with annual subsidies amounting to $279 billion. The United States is the world's top subsidizer at $502 billion, according to an International Monetary Fund report.[504] Domestic fuel subsidies in India reached 2 percent of GDP in 2011–2012. But you won't find these subsidies in the government's budget. "Fuel subsidies have been financed through a number of channels, including off-budget sources," according to the IMF report.

As long as the governments of China and India still support coal and provide regulatory protection and financial support to the coal industry, coal will prosper in the two most populous countries in the world. Governments apart from China and India have multinational financial organizations with their own agendas. Japan's Bank of International Cooperation has provided more than $10 billion to fund coal projects.[505]

These countries are certainly not funding coal on the basis of credible business plans. Fossil-fuel costs are extremely volatile in the short and medium terms and historically go up in in the long term.

Oil prices rose by more than 14 times in less than a decade (between 1999 and 2008). Natural gas prices are similarly volatile. Coal is no exception to this "expensive and volatile" rule.

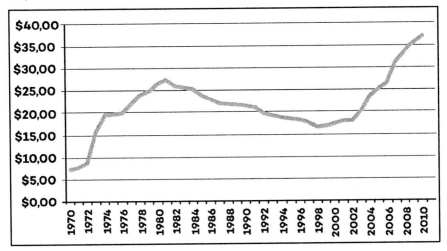

*Figure 10.4—Coal prices in the U.S. in USD per short ton. Prices are F.O.B. and do not include freight, shipping, or insurance. (Source: U.S. Energy Information Agency)*[506]

According to the U.S. Energy Information Agency, the F.O.B. price of a short ton of coal in the U.S. rose from $6.34 in 1970 to $36.91 in 2011 (see Figure 10.4). (F.O.B., or free on board, means the price at the mine, which doesn't include the cost of insuring or transporting coal, whether in the U.S. or overseas.) Coal prices have gone up by 5.8 times since 1970. In the meantime, the cost of solar PV has dropped by 154 times. Solar has improved its cost position relative to coal by nearly nine hundred times since 1970.

Coal prices have gone up despite production gains and increased market share since the 1970s. The decreased prices in the period 1980–2000 (see Figure 10.4) correspond roughly to productivity gains and massive layoffs in the industry. The coal mining industry increased its production by 21 percent from 1980 to 2000 while letting go of 69 percent of its workforce. The number of coal mine employees fell from 228,569 in 1980 to 71,522 in 2000 (see Figure 10.5).[507] Since 2000, coal prices have regained the upward trend as productivity gains have hit a plateau and even reversed course.

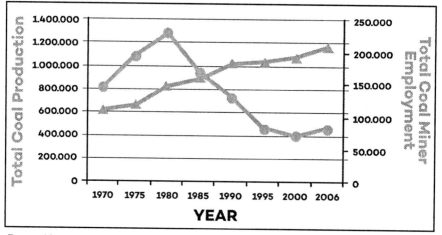

*Figure 10.5 – Total coal production vs. coal miner employment in the U.S. (Data Source: SourceWatch)[508]*

Where are coal prices headed? As with all fossil fuels, coal prices trend up. Short-term productivity gains may give an industry the opportunity to publicly declare a "new era" of "cheap power." The media, politicians, regulators, and others in the echo chamber then convince the public of the coming energy paradise brought by the dirty resource-based energy du jour (coal, gas, oil, nukes). But the evidence is clear: Conventional energy prices historically have gone up.

Again, solar PV will have gone down by a factor of four hundred, from $100/W in 1970 to 25c/W in 2020. By 2020, even assuming that coal prices stabilize at current prices (which is hard to believe), solar will have improved its cost position relative to coal by about 2,700 times.

The coal industry benefits when the government permits the taxpayer to pay for most of its costs. By way of government regulations and laws, the coal industry also benefits from cheap (or free) land; the ability to pollute air, water, and land at will; and the ability to renege on the pension and health care costs of its workers.[509]

## China: Water for Coal, not Food

For two decades, economists and so-called China experts and pundits have argued endlessly that China's yearly GDP growth is not sustainable. Year after year China has proved them all wrong. China has surpassed the United States as the largest auto market and the largest energy market in the world, both of which were unthinkable just a decade ago.

China has seemingly mastered the art and science of combining central planning and market-based entrepreneurship. If it can be manufactured, China will manufacture it. If it can be built, China will build it. There's seemingly nothing that it can't do. Or is there?

The one missing element in this equation is one of the most precious elements on earth: water. Cheap and abundant water has been as important to building civilizations as cheap and abundant energy. We need water to generate energy and we need energy to pump, transport, clean, and process water. Both water and energy are needed to grow food.

Lack of water is a limiting factor for fossil fuels and nuclear energy. About 15 percent of the world's freshwater withdrawal goes for energy, according to the World Bank.[510] Nearly half the world's population will live in areas of "high water stress" affecting energy and food insecurity.

China is already in a water crisis. China has 20 percent of the world's population but only 7 percent of its freshwater. The fast growth of its industry and population has caused the country to draw unsustainably on its rivers and aquifers. Since the 1950s, China has lost 27,000 of its 50,000 rivers.[511] The numbers that speak to China's water crisis are telling: 400 out of 600 cities, including 30 of the largest 32 cities in China, face water shortages to varying degrees. Ninety percent of city groundwater sources are contaminated; 70 percent of rivers and lakes are polluted. Three-hundred million people in China don't have access to safe drinking water.[512]

Despite the current crisis, total water use is expected to increase from 599 billion $m^3$ (158 trillion gallons) per year in 2010 to 670 billion $m^3$ (177 trillion gallons) per year in 2020.[513] This water will not feed the country's increasingly wealthy population. Agriculture is expected to decrease water withdrawals from 62 percent of total national freshwater in 2010 to 54 percent in 2020. Where is this valuable freshwater going? To quench the coal industry's insatiable thirst.

China, the world's largest consumer of coal, is already feeling the stress caused by coal's unquenchable thirst for freshwater. The coal industry drinks 138 billion $m^3$ (36.5 trillion gallons) per year, or 23 percent of the nation's freshwater. This number is expected to grow to 188 billion $m^3$ (49.7 trillion gallons) by 2020, raising the coal industry's consumption of China's water to 28 percent.[514]

Here's another way to look at how the government in China protects the coal industry. From 2010 to 2020, China will increase its freshwater consumption by 71.9 billion $m^3$ (19 trillion gallons). Of that increase, 49.9 billion $m^3$ (13.2 trillion gallons) will go to the coal sector. That is, 69 percent of the increase in valuable freshwater in China will go to the coal industry.

In China, coal truly is king, and like any king worth its salt, it doesn't have to pay for its drinking habit, clean after itself, or pay any of its expenses, however extravagant they might be.

The Chinese government support for coal is all the more vexing because coal is found in the most water-stressed parts of the country: the north and northwest. Sixty percent of new proposed coal plants in China are concentrated in just six provinces (Inner Mongolia, Shaanxi, Gansu, Ningxia, Shanxi, and Hebei) that together account for just 5 percent of the freshwater in China, according to the World Resources Institute.[515] Within those six provinces, 60 percent of coal plants are further concentrated in areas of "high or extremely high water stress" (see Figure 10.6).

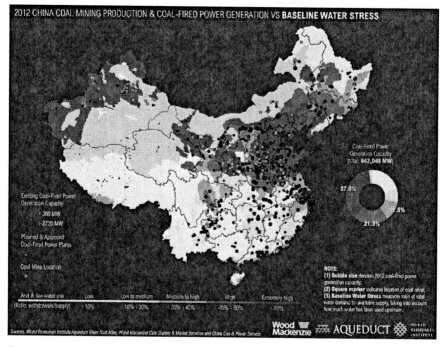

*Figure 10.6—Coal mining, power plants, and water stress in China. (Source: World Resources Institute)[516]*

To alleviate water concerns in the drying north, China is building some of the largest water projects in its history. The fitfully called "South North Water Transfer Project" is a $62–billion, multi-decade project to divert 44.8 billion $m^3$ (11.8 trillion gallons) of water from the Yangtze River basin in the south to the arid north.[517] Composed of three lines (eastern, central, and western routes) spanning thousands of miles of canals, tunnels, rivers, and reservoirs, this project is a major engineering undertaking.

The eastern route alone will be 716 miles (1,152 km) long and will be equipped with 23 pumping stations that will need 454 MW of power capacity (the production of a typical coal plant).[518]

If you look closely at the numbers, you'll notice the following:
- The coal sector, most of which is located in the north, is expected to increase its water demand by 49.9 billion m³ (13.2 trillion gallons) between 2010 and 2020.
- The "South North Water Transfer Project" is expected to deliver 44.8 billion m³ (11.8 trillion gallons) of water from the south to the north.

In other words, the "South-North Water Transfer Project" could be also known as the "Water for Coal Transfer Project."

The costs of building the South-North Water Transfer Project will not paid by the coal industry. The $62 billion water project is not directly reflected in the "cost" that they charge, but it is a real cost. As usual, the industry (and friendly governments) socialize their costs while privatizing their profits.

In the meantime, China has taken 8.5 million hectares (21 million acres) of farmland out of production since 1998.[519] The country also suffers from desertification on a grand scale. According to China's State Forestry Administration, 27 percent of the country (2.6 million square km, or 1 million square miles) now suffers from desertification.[520] Soil erosion impacts the lives of four-hundred million people and causes economic losses of $10 billion in China, according to the United Nations Convention to Combat Desertification (see Figure 10.7).[521]

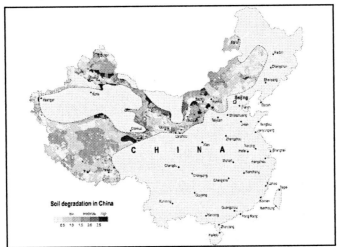

*Figure 10.7—Soil degradation in China. (Source: United Nations Convention to Combat Desertification)* [522]

China is literally sucking country the country dry to feed its coal industry.

Wen Jiabao, China's premier from 2003 to 2013, said that water shortages threaten "the very survival of the nation."[523] Can China's government afford to keep sacrificing its people for the benefit of its coal industry?

## Death by Coal

On October 22, 2013, the smog in northeastern China was so heavy the government closed roads, schools, and a major airport.[524] Visibility was as less than sixty feet (20 meters) in Harbin, a city of more than ten million people. All expressways in Heilongjiang province were closed.

The air was thick with a particulate matter 2.5, also known as PM2.5, which had reached a density of more than 1,000 micrograms per cubic meter ($\mu g/m^3$), or forty times the World Health Organization's maximum recommended level of 25 $\mu g/m^3$. Particulate matter 2.5 (PM2.5) is the result of burning coal and fossil fuels such as gasoline, diesel, and wood.

PM2.5 causes cardiopulmonary disease, cancer of the trachea, bronchus, and lung and acute respiratory infections.[525] It also leads to high plaque deposits in arteries, causing vascular inflammation and atherosclerosis, which can lead to heart attacks and other cardiovascular problems. Even short-term exposure to PM2.5 at elevated concentrations can cause heart disease.

According to the San Francisco Bay Area Air Quality Management District, in a population the size of the Bay Area (7.2 million in 2010), reducing ambient PM2.5 by just 1 $\mu g/m^3$ can save about 11,530 work days that would otherwise be lost. The level in Harbin, China was 1,000 $\mu g/m^3$.[526]

Air pollution in China is a human catastrophe. Outdoor air pollution caused 1.2 million premature deaths in China, according to The Lancet, a leading medical journal.[527]

Within China, the differences in life expectancy with respect to air pollution are stark. Life expectancy in northern China was 5.5 years less than southern China due to heart and lung disease caused by coal burning, according to a recent study published by the Proceedings of the National Academy of Sciences (PNAS).[528]

Much of the conversation about coal has to do with the damaging long-term effects of climate change, but coal already is one of the chief causes of death and disease around the world.

Coal is not cheap. We are paying for it with extra trips to the hospital, loss of lives, lower economic output, and loss of quality of life. The coal industry gets the profits and people pay for the costs.

India has a growing coal-induced human catastrophe of its own. Outdoor air pollution caused 600,000 deaths in India in 2010.[529] The country has about 120GW of coal today but has proposed building 519 GW of additional coal power plants. India's coal deposits are mainly composed of lignite, also known as brown coal.[530] Lignite is considered the lowest-quality coal, according to industry classifications.[531] Lignite has up to ten times the number of volatile organic compounds as U.S. anthracite. Because lignite has a lower heat content, more needs to be burned to get the same amount of energy as anthracite.[532]

If the additional 519 GW of coal power capacity is brought online in India, the country will quadruple its coal capacity infrastructure using lignite, the lowest-quality coal. Lignite generates ten times the particulate matter of anthracite coal. The human death toll in India because of coal pollution could be an order of magnitude higher than it is today, as many as six million deaths per year. That the Indian government knowingly supports this scheme is beyond rational comprehension.

Moreover, the water needs of the coal industry would devastate a country already ravaged by water mismanagement and desertification. Indian reports on soil degradation have increased by a factor of six, according to the United Nations Convention to Combat Desertification.[533]

Despite the millions of deaths caused by coal air pollution, government agencies that propose increases in coal production do so because, they say, "coal is cheap." Is it really?

American lives are also affected by coal pollution. According to the American Lung Association, coal pollution causes more than 24,000 early deaths in America through the toxic effect of coal on the lungs and other parts of the body.[534] To put this number in perspective, consider this: from 2001 to 2013, 5,281 American soldiers died in combat.[535] During those thirteen years, more than 312,000 Americans died because of coal pollution.

Coal costs the U.S. $500 billion per year in health, economic, and environmental damage, according to a Harvard University report.[536] That is, each man, woman, and child in America pays more than $1,600 per year for damages caused by the mining, transporting, burning, and disposing of coal. That's a massive tax. If the coal industry paid for the external damages it causes, it would have to pay us 26.89 ¢/kWh.[537] In other words, U.S. taxpayers are subsidizing the coal industry to the tune of 26.89 ¢/kWh.

The coal industry would not exist if there were a free market for energy and this industry was not protected by the government. In a free market, companies would not be able to unload the costs of pollution onto the people with the blessing and protection of their government.

When is the destruction of human life going to catch up with coal? When are government regulators going to stop aiding and abetting the coal industry in the killing of millions of human beings?

## The Final Disruption of Coal

In the fall of 2013, I taught a two-day technology innovation strategy course to some senior executives from Beijing. It was a crisp sunny day at Stanford University. During our lunch break, I went with my class for a photo-op at the Quad (see Figure 10.8). I never tire of admiring the Rodin sculpture garden at Stanford, the Memorial Church, and the way that the arches align with their shadows. I noticed several students looking up and taking pictures of the blue sky. I asked what they found interesting about the sky and they answered, "We never get a blue sky in China. Never. It is so beautiful!"

My response was "You will. By 2030." I teach disruption so they just smiled and said in unison "We hope so."

*Figure 10.8—Stanford University Quad. (Photo: Tony Seba)*

It's not that I expect the continental-scale Rube Goldberg engineering projects that support the coal industry to end soon. I just expect coal power plants to become stranded assets.

Lobbyists and their friendly regulators will continue to weave stories and spread misinformation to protect the coal industry. However, when unsubsidized solar becomes cheaper than subsidized coal, it will be hard for politicians and regulators to weave a convincing narrative about the benefits of coal. It will be hard for citizens to give their lives and turn over their wallets to support the coal industry when a cheaper, cleaner alternative exists.

Power utilities still tell the same old story about electricity being a "natural monopoly." Vertically integrated utility monopolies are accustomed to a cost-plus business model in which they pass on any increase in costs to the rate payers. Utilities love capital-intensive power plants like coal and nuclear with rising fuel costs because these power plants increase the utilities' income stream every year.

Solar is disruptive to that business model. Solar prices go down, not up. Solar can be installed on the customer's rooftop without any help from the utility. Solar needs no fuel to be mined, processed, transported, burned, or dumped back into the ground.

Coal prices have increased by 5.8 times since 1970 (see Figure 10.4). Coal prices are not just increasing; they are volatile. Coal companies themselves are well aware of the volatility in coal prices and are cashing in on it while they can. In the second quarter of 2013, Cloud Peak Energy made more money playing with financial derivatives than selling coal.[538] Energy flow is cash flow. Energy volatility is also cash flow. Coal companies will make money coming and going. But they can't guarantee you the same price for twenty years like solar and wind can.

The good news is that, despite massive protection and subsidies to the coal industry, coal is already being beat by solar and wind. According to models compiled by Bloomberg, a typical new coal plant will produce power for 12.8 ¢/kWh.[539] Solar is already beating this cost. First Solar's 50 MW Macho Springs project will sell solar energy to El Paso Electric for 5.79 ¢/kWh.[540] New Mexico's rate payers will pay for solar less than half what they would have paid for coal. Better yet, they will pay this low price for the twenty-year length of the contract.[541]

In California, the solar industry is well on its way to disrupting the existing electric utility paradigm. Danny Kennedy, co-founder Sungevity, a solar installer, told me recently, "Ninety-plus percent of our customers save money on day one."

Silicon Valley has built whole industries on the exponential improvement laws (such as Moore's Law) that govern information technology. For decades, Silicon Valley companies have relentlessly and consistently cut the cost of all aspects of computing. Today's smartphone has more computing power than yesterday's super-computer. The cost of solar, like the cost of computing, goes down relentlessly while the quality of solar goes up.

The same people who brought you Apple, Intel, and Google are now bringing you solar electricity. The bit disrupted the atom. Next, the bit plus the electron will obliterate atom-based utilities. The Industrial energy era is giving way to the knowledge-based energy era. It's that simple. Rube Goldberg energy engineering is no longer needed (see Figure 10.9).

*Figure 10.9—A Rube Goldberg machine at Google headquarters in Mountain View, California. (Machine built by Christopher Westhoff. Photo: Tony Seba.)*[542]

By 2020, the cost of solar panels will drop by two-thirds. That is, unsubsidized solar will be far cheaper than subsidized coal. Even residential solar costs will be lower than wholesale coal-generation costs.

The expected total cost of building a large solar power plant in 2020 will be $0.65 per Watt, according to GTM and Citibank.[543] This number won't be hard to achieve. In fact, we're almost there. As I've mentioned in this book, residential solar already costs $1.40/W in Australia. Large power plants have much lower costs than residential. Add the solar learning curve over the next six years and it would not be surprising if the total installed cost of a large solar plant were under $0.50/W by the year 2020.

Assuming Citibanks's estimate of $0.65/W to build a large solar plant by 2020 is correct, building a solar power plant near Los Angeles and financing it at a 4-percent cost of capital would get you a plant that produces electricity at 3.4 ¢/kWh. This levelized cost of electricity (LCOE) includes insurance, operations, maintenance and decommissioning. No unsubsidized dirty energy source on earth can produce power at this low cost. Not nuclear. Not coal. Not natural gas. Not oil. Not now. Not in 2020. Certainly not in 2030.

The coal industry would not exist without massive support from governments and citizens picking up most of its costs. Even within a captured regulatory environment, coal cannot compete with a technology cost curve like that of solar. Against coal, solar has improved its cost position by nine-hundred times and will improve that advantage by at least 2,700 times by 2020.

The only source of energy that will beat a solar plant built in 2020 is a solar plant built after 2020. That's because solar costs will keep decreasing well into the future. After the 20-year mortgage is paid off, the panels will continue to generate solar power at a cost of essentially zero. New solar panels are usually guaranteed for 20 years, but due to improvements in quality, they are expected to work for many decades after the initial two. Production will likely decrease (say, by 1 percent per year), but a 3 kW solar plant that is built in 2014 will likely produce 2.4kW twenty years later, in 2034, for zero cents per kWh. The panels will likely still be good for 2.2kW in 2044 and 2kW in 2054, again for zero cents per kWh.

Like the cost of computing technology, the cost of solar energy will keep going down for as far as the eye can see. We will experience something the world has never experienced before: first energy cost deflation, and then abundant, participatory, clean energy. And, oh yes, the sun will shine again in Beijing and this city will get its blue skies back.

Coal is obsolete. Look for investments in coal to become stranded assets.

## About the Author

Tony Seba is a lecturer in entrepreneurship, disruption, and clean energy at Stanford University. His courses include "Anticipating and Leading Market Disruption," "Clean Energy and Clean Transportation — Market and Investment Opportunities," and "Finance for Entrepreneurs."

Tony Seba is also the author of Solar Trillions – 7 Market and Investment Opportunities in the Emerging Clean Energy Economy and Winners Take All – 9 Fundamental Rules of High Tech Strategy.

Tony Seba is currently advising companies developing more than 400 MW of solar power and wind power plants globally. He has also advised venture investors on high-technology company investments. He is an advisor to solar accelerator SFUN Cube.

He is a Silicon Valley entrepreneur and executive with more than twenty years of experience in disruptive, fast-growth technology businesses. He was an early employee of the Internet powerhouse Cisco Systems and the Internet security standard-setter RSA Data Security. As director of strategic planning at RSA Data Security, he helped the company with new product and market creation. He also helped guide the company's $200-million merger with Security Dynamics. He was the co-founder, president, and CEO of PrintNation.com, for which he raised more than $31 million in venture funding. He established PrintNation.com as the undisputed leader in its market segment and disrupted this $100 billion industry. The company won numerous "best of" awards and was listed in the Upside Hot 100 and the Forbes.com B2B "Best of the Web."

Tony Seba is a recognized thought leader and keynote speaker in entrepreneurship, disruption, and the future of energy and transportation. His speaking clients include Google, the California League of Cities, the Inter-American Development Bank, and the Institute for The Future.

His leadership has been recognized in publications such as Investors Business Daily, Business Week, Upside, and Success. He has served on many boards of directors and advisory boards. He holds a BS in Computer Science and Engineering from Massachusetts Institute of Technology (MIT) and an MBA from the Stanford University Graduate School of Business.

Follow Tony Seba at his website www.tonyseba.com and his Twitter account @tonyseba

# Endnotes

[1]"Exploring the Role Ecosystems in Evolving Cloud Markets", IBM Smart Cloud White Paper, 2012. http://www.ibm.com/midmarket/lk/en/att/pdf/lk_en_White_Paper_Cloud_2012.pdf

[2] Bajarin, Ben, "The State of Mobile Technology", Time, March 5, 2012. http://techland.time.com/2012/03/05/the-state-of-mobile-technology/

[3]"Constine, Josh, "40% of YouTube traffic Now Mobile", TechCrunch.com, Oct 17, 2013. http://techcrunch.com/2013/10/17/youtube-goes-mobile/

[4]"Explaining the NOAA Sea Level Rise Viewer", Climate.Gov, Oct 29, 2013 http://www.climate.gov/news-features/videos/explaining-noaa-sea-level-rise-viewer

[5]J. Alstan Jakubiec and Christoph F. Reihnart, "Towards Validated Urban Photovoltaic Potential and Solar Radiation Maps based on Lidar measurements, GIS data, and hourly days in simulations", Building Technology Program, Massachusetts Institute of Technology, Cambridge, MA

[6]Edelstein, Stephen, "Nissan Shows Fully Autonomous Cars it plans to Build by 2020", GreenCarReports.com, Sept 9, 2013. http://www.greencarreports.com/news/1086566_nissan-shows-fully-autonomous-cars-it-plans-to-build-by-2020

[7]"TSensor Summit", Stanford University, http://www.tsensorssummit.org/

[8]Loder, Asjvlyn, "U.S. Shale-Oil Boom May Not Last as Fracking Wells Lack Staying Power", Business Week, Oct 10, 2013. http://www.businessweek.com/articles/2013-10-10/u-dot-s-dot-shale-oil-boom-may-not-last-as-fracking-wells-lack-staying-power

[9]Lavrinc, Damon, "In Automotive First, Tesla does Over-The-Air Software Patch", Wired.com, Sept 24, 2012. http://www.wired.com/autopia/2012/09/tesla-over-the-air/

[10]"Kumparak, Greg, "Live from iPhone 5S Announcement", TechCrunch, Sept 10, 2013, http://techcrunch.com/2013/09/10/live-blog-from-apples-iphone-5s-announcement/

[11]"Enough Copper for a Hundred Years", Materials Today, July 4, 2013, http://www.materialstoday.com/metals-alloys/news/enough-copper-for-a-hundred-years/

[12]"oossens, Ehren, and Martin, Christopher, "First Solar: May Sell Solar at Less than Coal", Bloomberg News, February 1, 2013.
http://www.bloomberg.com/news/2013-02-01/first-solar-may-sell-cheapest-solar-power-less-than-coal.html

[13]"U.S. Solar Market Insight Report, 2013 Year In Review", Solar Energy Industries Association, http://www.seia.org/research-resources/solar-market-insight-report-2013-year-review

[14] Sherwood, Larry, "Solar Market Trends 2011", Interstate Renewable Energy Council, August 2012, http://www.irecusa.org/wp-content/uploads/IRECSolarMarketTrends-2012-Web-8-28-12.pdf

[15] Seba, Tony, "Will Germany Achieve 100% Solar Power by 2020?", August 6, 2012 http://tonyseba.com/cleanenergyeconomy/germany-100-solar-power-by-2020/

[16] Meikle, Brad, "Meikle Capital, Technology Equilibrium Fund, LP", July 8, 2013 newsletter

[17] Meikle, Brad, "Meikle Capital, Technology Equilibrium Fund, LP", July 8, 2013 newsletter

[18] Meikle, Brad, "Meikle Capital, Technology Equilibrium Fund, LP", July 8, 2013 newsletter

[19] "German solar PV, wind peak at 59.1% of electricity production on October 3rd, 2013", SolarServer.com, Oct 7, 2013, http://www.solarserver.com/solar-magazine/solar-news/current/2013/kw41/german-solar-pv-wind-peak-at-591-of-electricity-production-on-october-3rd-2013.html

[20] Morris, Craig, "Denmark Surpasses 100 percent Wind Power", Energy Transition, Nov 8, 2013. http://energytransition.de/2013/11/denmark-surpasses-100-percent-wind-power/

[22] "Germany's Solar Power Systems Set New Solar Record (Germany Crushing US in Solar Power)", Cost Of Solar, Nov 2013. http://costofsolar.com/germany-solar-power-systems/

[23] Davis, Tina, and Goossens, Ehren, "Buffett Utility Buys $2.5 Billion SunPower Solar Projects", January 2, 2013. http://www.bloomberg.com/news/2013-01-02/buffett-utility-buys-sunpower-projects-for-2-billion.html

[24] MidAmerican Energy, US SEC Form10K for fiscal year ended December 31, 2012, http://www.midamerican.com/include/pdf/sec/20121231_79_mec_annual.pdf

[25] "$1 Billion Bond Offering Completed for World's Largest Solar Project", SustainableBusiness.com, June 28, 2013. http://www.sustainablebusiness.com/index.cfm/go/news.display/id/25018

[26] Shahan, Zachary, "Shell Bullish on Solar Despite Dropping Solar", SolarLove.org, March 3, 2013, http://solarlove.org/shell-bullish-on-solar-despite-dropping-solar-but-much-more-in-its-new-scenarios-than-that/

[27] "Global Market Outlook 2013-2017", EPIA European Photovoltaic Industry Association. http://www.epia.org/fileadmin/user_upload/Publications/GMO_2013_-_Final_PDF.pdf

[28] Meza, Edgar, "China's Solar Capacity to Reach 10 GW in 2013", PV Magazine, December 4, 2013. http://www.pv-magazine.com/news/details/beitrag/chinas-solar-capacity-to-reach-10-gw-in-2013_100013650/

[29] Wikipedia contributors, "Solar Power in Germany", Wikipedia, the Free Encyclopedia. http://en.wikipedia.org/wiki/Solar_power_in_Germany

[30] Wikipedia contributors, "Solar Power in Germany", Wikipedia, the Free Encyclopedia. http://en.wikipedia.org/wiki/Solar_power_in_Germany

[31] "National Transmission Development Plan for the National Electricity Market", Australia Market Operator, December 2013, AEMO Australian Energy Market Operator, ABN 94 072 010 327. http://www.pennenergy.com/content/dam/Pennenergy/online-articles/2013/December/2013_NTNDP.pdf.pdf

[32]" "Wind In Power – 2011 European Statistics", The European Wind Energy Association, Feb 2012. http://www.ewea.org/fileadmin/files/library/publications/statistics/Wind_in_power_2011_European_statistics.pdf

[33] Platzer, Michaela D, "U.S. Solar Photovoltaic Manufacturing: Industry Trends, Global Competition, Federal Support", Congressional Research Service, June 13, 2012, R42509, www.crs.gov

[34] "ReConsidering the Economics of PV Power", Bloomberg New Energy Finance, 2012

[35] Petroleum and other Liquids", US Energy Information Agency. http://www.eia.gov/dnav/pet/hist/LeafHandler.ashx?n=PET&s=F000000__3&f=A

[36] "1970 Economy / Prices", 1970sFlashback.com. http://www.1970sflashback.com/1970/economy.asp

[37] Kind, Peter, "Disruptive Challenges: Financial Implications and Strategic Response to a Changing Retail Electric Business", Edison Electric Institute, January 2013

[38] Linbaugh, Kate, and Kell, John, "GE Ends Solar-Panel Push, Sells Technology to First Solar", Wall Street Journal, August 6 2013. http://online.wsj.com/news/articles/SB10001424127887323514404578652533484101340

[39] Montaigne, Fen, "A Power Company President Ties His Future to Green Energy", Yale Environment 360, November 9, 2011. http://e360.yale.edu/feature/solar_power_nrg_president_crane_ties_future_to_renewable_energy/2462/

[40] "The Impact of Local Permitting on the Cost of Solar Power", SunRun, January, 2011. http:www.sunrunhome.com/download_file/view/414/189

[41] Rinaldi, Nicholas, "Solar PV Modules Costs to Fall to 36 cents per watt by 2017", GreentechMedia,,June 18, 2013. http://www.greentechmedia.com/articles/read/solar-pv-module-costs-to-fall-to-36-cents-per-watt

[42] "Statistic data on the German Solar power (photovoltaic) industry", German Solar Industry Association (BSW-Solar), April 2014. http://www.solarwirtschaft.de/fileadmin/media/pdf/2013_2_BSW-Solar_fact_sheet_solar_power.pdf

[43] Martin, James, "Solar PV Price Index–July 2013", Solar Choice, July 4, 2013. http://www.solarchoice.net.au/blog/solar-pv-price-index-july-2013/

[44] Seel, Joachim, Barbose Galen, and Wiser, Ryan, "Why Are Residential PV Prices in Germany So Much Lower Than in the United States? A Scoping Analysis", Lawrence Berkeley National Laboratory, September 2012. http://rael.berkeley.edu/sites/default/files/lbnl-german-price-presentation-final.pdf

[45] Martin, James, "Solar PV Price Index–July 2013", Solar Choice, July 4, 2013. http://www.solarchoice.net.au/blog/solar-pv-price-index-july-2013/

[46] Drury, Easan, Denhom, Paul, and Margolis, Robert, "Sensitivity of Rooftop PV Projections in the SunShot Vision Study to Market Assumptions", National Renewable Energy Lab, NREL/TP-6A20-54620,January 2013

[47] Woody, Todd, "First Solar Shares Spike As It Pursues Big Solar Without Subsidies Strategy", Forbes.com, July 2, 2012. http://www.forbes.com/sites/toddwoody/2012/08/02/first-solar-shares-spike-as-it-pursues-big-solar-without-subsidies-strategy/

[48] Rinaldi, Nicholas, "Solar PV Modules Costs to Fall to 36 cents per watt by 2017", GreentechMedia,,June 18, 2013. http://www.greentechmedia.com/articles/read/solar-pv-module-costs-to-fall-to-36-cents-per-watt

[49] Channell, Jason, Lam, Timothy, and Pourreza, Shahriar, "Shale & Renewables: A Symbiotic Relationship", Citi Research, Sept 12, 2012

[50] Channell, Jason, Lam, Timothy, and Pourreza, Shahriar, "Shale & Renewables: A Symbiotic Relationship", Citi Research, Sept 12, 2012

[51] "Historical California Electricity Demand", California Energy Commission, CA Energy Almanac. http://energyalmanac.ca.gov/electricity/historic_peak_demand.html

[52] ReneSola, 156 Series Monocrystalline Solar Module, Irradiation efficiencies vary from 15.9% to 16.2%, available at http://www.renesola.com

[53] Martin, James, "Solar PV Price Index–July 2013", Solar Choice, July 4, 2013. http://www.solarchoice.net.au/blog/solar-pv-price-index-july-2013/

[54] Martin, James, "Solar PV Price Index–July 2013", Solar Choice, July 4, 2013. http://www.solarchoice.net.au/blog/solar-pv-price-index-july-2013/

[55] Scott, Brandon, "Powering the Future – Small City has Big Solar Goals", CBS News, August 19, 2013. http://www.cbsnews.com/news/powering-the-future-small-city-has-big-solar-goals/

[56]Trabish, Herman, "Lancaster, CA Becomes First US City to Require Solar", March 27, 2013. http://www.greentechmedia.com/articles/read/Lancaster-CA-Becomes-First-US-City-to-Require-Solar

[57]Tim Flannery, Tim, and Sahajwalla, Veena, "The Critical Decade: Australia's Future – Solar Energy", Climate Commission, 2013. http://www.climatecouncil.org.au/uploads/497bcd1f058be45028e3df9d020ed561.pdf

[58] Frank M Bass et al, "DIRECTV - Forecasting the Diffusion of a New Technology Prior to Product Launch", INTERFACES 31: 3, Part 2 of 2, May–June 2001 (pp. S82–S93)

[59] Frank M Bass et al, "DIRECTV - Forecasting the Diffusion of a New Technology Prior to Product Launch", INTERFACES 31: 3, Part 2 of 2, May–June 2001 (pp. S82–S93)

[60] Wikipedia contributors, "Potassium Nitrate", Wikipedia, the Free Encyclopedia. http://en.wikipedia.org/wiki/Potassium_nitrate

[61] Wikipedia contributors, "Capacity Factor", Wikipedia, the Free Encyclopedia. http://en.wikipedia.org/wiki/Capacity_factor, as of June 20, 2011.

[62] Maloney, Michael T., "Analysis of Load Factors at Nuclear Power Plants", http://works.bepress.com/cgi/viewcontent.cgi?article=1009&context=michael_t_maloney , retrieved June 20, 2011.

[63] "Major Solar Projects in the United States Operating, Under Construction, or Under Development", Solar Energy Industries Association, November 26, 2013

[64] Wikipedia contributors, "World Wide Web", Wikipedia, the Free Encyclopedia. http://en.wikipedia.org/wiki/World_Wide_Web#Web_servers

[65] Deken, Jean Marie, "The Early World Wide Web at SLAC", Stanford Linear Accelerator Center, May 31, 2006. http://www.slac.stanford.edu/history/earlyweb/history.shtml

[66] Lomas, Natasha, "10BN+ Wirelessly Connected Devices Today, 30BN+ In 2020,s 'Internet Of Everything', Says ABI Research", TechCrunch, May 9, 2013. http://techcrunch.com/2013/05/09/internet-of-everything/

[67] "National Transmission Development Plan for the National Electricity Market", Australia Market Operator, December 2013, AEMO Australian Energy Market Operator, ABN 94 072 010 327. http://www.pennenergy.com/content/dam/Pennenergy/online-articles/2013/December/2013_NTNDP.pdf.pdf

[68] Wang, Yue, "More People Have Cell Phones Than Toilets, U.N. Study Shows", Time, March 25, 2013. http://newsfeed.time.com/2013/03/25/more-people-have-cell-phones-than-toilets-u-n-study-shows/

[69] Bells, Mary, "The History of Plumbing: Toilets", About.com. http://inventors.about.com/od/pstartinventions/a/Plumbing_3.htm

[70] Kenney, Kim, "Cars in the 1920's", Suite 101. http://suite101.com/article/cars-in-the-1920s-a90169

[71] Trabish, Herman, "Sunrun Closes $630M in Rooftop Solar Funds From JPMorgan, US Bank", GreentechMedia, June 26, 2013. http://www.greentechmedia.com/articles/read/Sunrun-Closes-630-Million-in-Rooftop-Solar-Funding-from-JPMorgan-US-Bank

[72] Mims, Christopher et al, "World Changing Ideas: 20 Ways to Build a Cleaner, Healthier, Smarter World", Scientific American, December 2009. http://www.scientificamerican.com/article/world-changing-ideas/

[73] Trabish, Herman, "Sunrun Closes $630M in Rooftop Solar Funds From JPMorgan, US Bank", GreentechMedia, June 26, 2013. http://www.greentechmedia.com/articles/read/Sunrun-Closes-630-Million-in-Rooftop-Solar-Funding-from-JPMorgan-US-Bank

[74] "Vivint Solar Raises $540 million in Residential Solar", PV Magazine, Oct 21, 2013. http://www.pv-magazine.com/news/details/beitrag/vivint-raises-540-million-for-residential-solar_100013085/

[75] Hoium, Travis, "SolarCity's Growth Binge Continues", The Motley Fool, August 8, 2013. http://www.fool.com/investing/general/2013/08/08/solarcitys-growth-binge-continues.aspx

[76] SolarCity Corporation, Yahoo! Finance, http://finance.yahoo.com/echarts?s=SCTY+Interactive#symbol=SCTY;range=1y

[77] Davidson, Paul, "Prices for rooftop solar systems fall as supply grows", USA Today, January 23, 2009. http://usatoday30.usatoday.com/money/industries/energy/environment/2009-01-12-solar-panels-glut_N.htm

[78] Berman, Jillian, "U.S. Median Annual Wage Falls To $26,364 As Pessimism Reaches 10-Year High", The Huffington Post, January 23, 2012. http://www.huffingtonpost.com/2011/10/20/us-incomes-falling-as-optimism-reaches-10-year-low_n_1022118.html

[79] Davis, Benjamin et al, "California Solar Cities 2012", Environment California Research & Policy Center, January 24, 2012. http://www.environmentcalifornia.org/reports/cae/californias-solar-cities-2012

[80] Wikipedia contributors, "PACE Financing", Wikipedia, the Free Encyclopedia, http://en.wikipedia.org/wiki/PACE_Financing

[81] "What is PACE?", PACENow. http://pacenow.org/blog/about-pace/

[82] "FHFA Statement on Certain Energy Retrofit Loan Programs", Federal Housing Finance Agency, July 6, 2010

[83] Quackenbush, Jeff, "Santa Rosa-based Ygrene leads $650 million green-retrofit effort", North Bay Business Journal, September 23, 2011. http://www.northbaybusinessjournal.com/40821/santa-rosa-based-ygrene-leads-650-million-green-retrofit-effort/

[84] "Plant A Seed For Solar Energy", Indiegogo. http://www.indiegogo.com/projects/plant-a-seed-for-solar-energy

[85] Morris, Craig, "Denmark Surpasses 100 percent Wind Power", Energy Transition, Nov 8, 2013. http://energytransition.de/2013/11/denmark-surpasses-100-percent-wind-power/

[86] Wikipedia contributors, "Wind Power in Denmark", Wikipedia, the Free Encyclopedia. http://en.wikipedia.org/wiki/Wind_power_in_Denmark

[87] Wikipedia contributors, "Wind Power in Denmark", Wikipedia, the Free Encyclopedia. http://en.wikipedia.org/wiki/Wind_power_in_Denmark

[88] Wikipedia contributors, "Wind Power in Denmark", Wikipedia, the Free Encyclopedia, retrieved 24 July 2013, http://en.wikipedia.org/wiki/Wind_power_in_Denmark

[89] Wikipedia contributors, "File: Wind in Denmark 1977-2011", Wikipedia, the Free Encyclopedia, http://en.wikipedia.org/wiki/File:Wind_in_Denmark_1977_2011_large.png

[90] Bayar, Tildy, "Dutch Wind Turbine Purchase Sets World Crowdfunding Record", Renewable Energy World, September 24, 2103. http://www.renewableenergyworld.com/rea/news/article/2013/09/dutch-wind-turbine-purchase-sets-world-crowdfunding-record

[91] WindCentrale company website: https://www.windcentrale.nl/faq/

[92] Wikipedia contributors, "Golden Gate Bridge", Wikipedia, the Free Encyclopedia
http://en.wikipedia.org/wiki/Golden_Gate_Bridge#History

[93] Wikipedia contributors, "Golden Gate Bridge, Highway and Transportation District",
Wikipedia, the Free Encyclopedia, http://en.wikipedia.org/wiki/Golden_Gate_Bridge,_
Highway_and_Transportation_District

[94] "Bond Measure Passes - Against the Odds?", Golden Gate Bridge, Highway and
Transportation District, http://goldengatebridge.org/research/BondMeasure.php

[95] "History of Golden Gate Ferry", Golden Gate Bridge, Highway and Transportation District,
http://goldengateferry.org/researchlibrary/history.php

[96] Mosaic Website, "Browse Investments", https://joinmosaic.com/browse-investments

[97] "Key Statistics", Federal Deposit Insurance Corporation, http://www2.fdic.gov/idasp/index.
asp

[98] Doom, Justin, and Buhayar, Noah, "Buffett Plans More Solar Bonds After Oversubscribed
Deal", Bloomberg, March 1, 2012.
http://www.bloomberg.com/news/2012-02-29/buffett-plans-more-solar-bonds-after-
oversubscribed-topaz-deal.html

[99] "Daily Treasury Yield Curve Rates", Resource Center, U.S. Department of the Treasury,
http://www.treasury.gov/resource-center/data-chart-center/interest-rates/Pages/TextView.
aspx?data=yield

[100] NASDAQ Composite (^IXIC), Yahoo! Finance, http://finance.yahoo.com/

[101] Hoff, Ivan, "26 Market Wisdoms from Warren Buffett", http://stocktwits50.
com/2013/09/09/26-market-wisdoms-from-warren-buffett/

[102] "SolarCity Announces Pricing of Securitization", SolarCity Press Release, November 13,
2013, http://investors.solarcity.com/releasedetail.cfm?ReleaseID=807221

[103] Chen, Xilu, and Chen, Weili, "SolarCity LMC Series I LLC (Series 2013-1)", Standard &
Poor's, RatingsDirect, November 11, 2013. http://www.standardandpoors.com/spf/upload/
Ratings_US/SolarCity_LMC_11_11_13.pdf

[104] Yoon, Al, "Surge in Asset-Backed Bond Sales", Wall Street Journal, September 10, 2012.
http://online.wsj.com/news/articles/SB10000872396390443921504577643931647213966

[105] Hannon Armstrong (HASI) Form 10Q for period ending March 31, 2013, from Yahoo! Finance, http://yahoo.brand.edgar-online.com/displayfilinginfo.aspx?FilingID=9316256-878-162223&type=sect&dcn=0001193125-13-233593

[106] Wikipedia contributors, "Real Estate Investment Trust", Wikipedia, the Free Encyclopedia, http://en.wikipedia.org/wiki/Real_estate_investment_trust

[107] Wikipedia contributors, "Real Estate Investment Trust", Wikipedia, the Free Encyclopedia, http://en.wikipedia.org/wiki/Real_estate_investment_trust

[108] "Statement of Dan W. Reicher Executive Director Steyer-Taylor Center for Energy Policy & Finance at Stanford University Professor, Stanford Law School Lecturer, Stanford Graduate School of Business to the House Committee on Oversight and Government Reform Subcommittee on Energy Policy, Health Care and Entitlements Hearing on Oversight of the Wind Energy Production Tax Credit", October 2, 2013, available at http://oversight.house.gov/wp-content/uploads/2013/10/Reicher.pdf

[109] Hannon Armstrong Sustainable Infrastructure Capital Inc (HASI), Morningstar website: http://quotes.morningstar.com/stock/hasi/s?t=HASI

[110] "Hannon Armstrong (HASI) Completes $100,000,000 Asset-Backed Securitization of 2.79% Sustainable Yield Bonds", press release, December 23, 2013, http://www.prnewswire.com/news-releases/hannon-armstrong-hasi-completes-100000000-asset-backed-securitization-of-279-sustainable-yield-bonds-236999811.html

[111] "Statement of Dan W. Reicher Executive Director Steyer-Taylor Center for Energy Policy & Finance at Stanford University Professor, Stanford Law School Lecturer, Stanford Graduate School of Business to the House Committee on Oversight and Government Reform Subcommittee on Energy Policy, Health Care and Entitlements Hearing on Oversight of the Wind Energy Production Tax Credit", October 2, 2013, available at http://oversight.house.gov/wp-content/uploads/2013/10/Reicher.pdf

[112] "Company History", Kinder Morgan company website: http://www.kindermorgan.com/about_us/kmi_history.cfm

[113] Wikipedia contributors, "Kinder Morgan", Wikipedia, the Free Encyclopedia, http://en.wikipedia.org/wiki/Kinder_Morgan

[114] "Rise of the Distorporation", The Economist, Oct 26, 2013. http://www.economist.com/news/briefing/21588379-mutation-way-companies-are-financed-and-managed-will-change-distribution

[115] "The Master Limited Partnerships Parity Act", United States Senator Chris Coons, http://www.coons.senate.gov/issues/master-limited-partnerships-parity-act

[116] Mormann, Felix, and Reicher, Dan, "How to Make Renewable Energy Competitive", New York Times, June 1, 2012. http://www.nytimes.com/2012/06/02/opinion/how-to-make-renewable-energy-competitive.html

[117] "Solar Investment Tax Credit (ITC)", Solar Energy Industries Association, http://www.seia.org/policy/finance-tax/solar-investment-tax-credit

[118] Weiss, Daniel, and Germain, Tiffany, "Big Oil, Big Profits, Big Tax Breaks", Center for American Progress, November 5, 2013, http://www.americanprogress.org/issues/green/news/2013/11/05/78807/big-oil-big-profits-big-tax-breaks/

[119] "S.795 - Master Limited Partnerships Parity Act", 113th Congress (2013-2014), http://beta.congress.gov/bill/113th/senate-bill/795/committees

[120] Natter, Ari, "Estimated to Cost $1.3 Billion Over 10 Years", Bloomberg Daily Tax Report, Nov 19, 2013. http://www.mw-cleantechcapital.com/files/2013/11/BNA-Tax-Report.pdf

[121] "SolarCity Announces Pricing of Securitization", SolarCity press release, November 13, 2013, http://investors.solarcity.com/releasedetail.cfm?ReleaseID=807221

[122] "About Clean Power Finance", company website: http://www.cleanpowerfinance.com/about-clean-power-finance/

[123] "Solar Investment Tax Credit (ITC)", Solar Energy Industries Association, http://www.seia.org/policy/finance-tax/solar-investment-tax-credit

[124] "Brannon Solar Renewable Energy Contract", City Council Staff Report, City of Palo Alto, November 5, 2012. http://www.cityofpaloalto.org/civicax/filebank/documents/31752

[125] "Brannon Solar Renewable Energy Contract", City Council Staff Report, City of Palo Alto, November 5, 2012. http://www.cityofpaloalto.org/civicax/filebank/documents/31752

[126] "Understand Your Electric Charges", PG&E company website, retrieved July 19, 2013, http://www.pge.com/myhome/myaccount/charges/

[127] "Understand Your Electric Charges", PG&E website retrieved July 19, 2013, http://www.pge.com/myhome/myaccount/charges/

[128] "Clean Energy Australia Report 2012", Clean Energy Council

[129] Wikipedia contributors, "Demographics of Australia", Wikipedia, the Free Encyclopedia, http://en.wikipedia.org/wiki/Demographics_of_Australia

[130] http://quickfacts.census.gov/qfd/states/06000.html

[131] Drew, Tim et al, California Solar Initiative, Annual Program Assessment Report, California Public Energy Commission, June 2013, http://www.cpuc.ca.gov/NR/rdonlyres/7A350E8E-3666-4AA5-98E3-5E9C812D3DE6/0/CASolarInitiativeCSIAnnualProgAssessmtJune2013FINAL.pdf

[132] "U.S. Solar Market Grows 76% in 2012", Solar Energy Industries Association, March 14, 2013, http://www.seia.org/news/us-solar-market-grows-76-2012-now-increasingly-competitive-energy-source-millions-americans

[133] Chan, Albert et al, "Australia Competing With Germany On Low Solar PV Prices", CleanTechnica, http://cleantechnica.com/2013/04/03/australia-competing-with-germany-on-low-solar-pv-prices/

[134] Grimes, John, "Australian PV Market Report", Australian Solar Council, presentation at Intersolar North America, July 2013

[135] Arizona Public Service, APC A.C.C. 5724, Rate Schedule ET-SP January, 2012

[136] "How to Lose Half a Trillion Euros", The Economist, October 10, 2013. http://www.economist.com/news/briefing/21587782-europes-electricity-providers-face-existential-threat-how-lose-half-trillion-euros

[137] Arizona Public Service, APC A.C.C. 5724, Rate Schedule ET-SP January, 2012

[138] "What is Peak Day Pricing?", Peak Energy Agriculture Rewards by Enernoc, Inc. http://pearcalifornia.com/what-is-peak-day-pricing

[139] "The Benefits of Uniform Clearing Price Auctions for Pricing Electricity – Why Pay-As-Bid Auctions Do not Cost Less", ISO New England, March 2006, http://www.iso-ne.com/pubs/whtpprs/uniform_clearing_price_auctions.pdf

[140] "Understanding the Markets - Clearing Price Auctions", NYISO, http://www.nyiso.com/public/about_nyiso/understanding_the_markets/clearing_price_auctions/index.jsp

[141] "How to Lose Half a Trillion Euros", The Economist, October 10, 2013, http://www.economist.com/news/briefing/21587782-europes-electricity-providers-face-existential-threat-how-lose-half-trillion-euros

[142] "Annual Energy Outlook 2013", U.S. Energy Information Administration, http://www.eia.gov/oiaf/aeo/tablebrowser/#release=AEO2013&subject=0-AEO2013&table=8-AEO2013&region=0-0&cases=ref2013-d102312a

[143] "Quarterly Report on European Electricity Markets", Volume 5, Issue 1: January 2012 – March 2012, European Commission Directory General for Energy

[144] "Capital Costs for Transmission and Substations", Recommendations for Western Electricity Coordinating Council (WECC) Transmission and Expansion Planning, B&V Project No. 176322, October 2012

[145] Silverstein, Alison, "TRANSMISSION 101", NCEP Transmission Technologies Workshop April 20-21, 2011

[146] "LIPA Announces Major New Plans for Additional Solar Energy for Long Island", Long Island Power Authority press release, July 12, 2013. http://www.lipower.org/newscenter/pr/2013/071213-solar.html

[147] "Brannon Solar Renewable Energy Contract", City Council Staff Report, City of Palo Alto, November 5, 2012. http://www.cityofpaloalto.org/civicax/filebank/documents/31752

[148] "Solar Means Business – Top U.S. Commercial Solar Users", Solar Energy Industries Association and VoteSolar, Oct 2013.

[149] "Walmart Announces 10 new solar installations in Maryland", Walmart company press release, June 25, 2013, http://news.walmart.com/news-archive/2013/06/25/walmart-announces-10-new-solar-installations-in-maryland

[150] "IKEA Expanding Rooftop Array on Maryland Distribution Center to 4.9 MW", SolarIndustry.com, November 7, 2013

[151] "Walmart Announces 10 new solar installations in Maryland", Walmart company press release, June 25, 2013, http://news.walmart.com/news-archive/2013/06/25/walmart-announces-10-new-solar-installations-in-maryland

[152] "The New Energy Consumer Handbook", Accenture, June 2013, http://nstore.accenture.com/acn_com/PDF/Accenture-New-Energy-Consumer-Handbook-2013.pdf

[153] Clover, Ian, "VW Builds Car Industry's Largest Solar Installation In Spain", PV Magazine, November 29, 2031, http://www.pv-magazine.com/news/details/beitrag/vw-builds-car-industrys-largest-solar-installation-in-spain_100013607/

[154] "Environmental Responsibility", Apple company website, http://www.apple.com/environment/renewable-energy/

[155] "Prologis Profile – Americas 2Q 2013", company presentation, http://www.prologis.com/docs/Prologis_Profile_Americas_2Q13.pdf

[156] weet, Cassandra, "NRG, Prologis, Embark on Solar Rooftop Project", Wall Street Journal, June 22, 2011, http://online.wsj.com/news/articles/SB10001424052702304791204576401883873490842

[157] Solar Means Business – Top U.S. Commercial Solar Users", Solar Energy Industries Association and VoteSolar, Oct 2013.

[158] Wikipedia contributors, "Yoky Matsuoka", Wikipedia, the Free Encyclopedia, http://en.wikipedia.org/wiki/Yoky_Matsuoka

[159] "Heating and Cooling", US Department of Energy, http://energy.gov/public-services/homes/heating-cooling

[160] Fischer, Barry, "Hot and heavy energy usage: How the demand and price for electricity skyrocketed on a 100° day", oPower.com, Sept 5, 2012
http://blog.opower.com/2012/09/hot-and-heavy-energy-usage-how-the-demand-and-price-for-electricity-skyrocketed-on-a-100-day/

[161] "Texas Heat Wave, August 2011: Nature and Effects of an Electricity Supply Shortage", U.S. Energy Information Agency, http://www.eia.gov/todayinenergy/detail.cfm?id=3010

[162] "Hot and heavy energy usage: How the demand and price for electricity skyrocketed on a 100° day", oPower.com, Sept 5, 2012
http://blog.opower.com/2012/09/hot-and-heavy-energy-usage-how-the-demand-and-price-for-electricity-skyrocketed-on-a-100-day/

[163] "National Overview - June 2012", National Climatic Data Center, National Oceanic and Atmospheric Administration, http://www.ncdc.noaa.gov/sotc/national/2012/6

[164] "What You May Not Know About Galaxy S4 Innovative Technology", April 10, 2013, Samsung company website, http://global.samsungtomorrow.com/?p=23610

[165] Upbin, Bruce, "Monsanto Buys Climate Corp for $930 million", Forbes.com, October 2, 2013, http://www.forbes.com/sites/bruceupbin/2013/10/02/monsanto-buys-climate-corp-for-930-million/

[166] Rothstein, Edward, "An Emphasis on Newton's Laws (and a Little Lawlessness) - The New Exploratorium Opens in San Francisco", New York Times, April 16th, 2013, http://www.nytimes.com/2013/04/17/arts/design/the-new-exploratorium-opens-in-san-francisco.html

[167] Woolsey, Christina, "Sustain – The Museum as Exhibit", San Francisco: Exploratorium, November 2012

[168] Woolsey, Christina, "Sustain – The Museum as Exhibit", San Francisco: Exploratorium, November 2012

[169] Kind, Peter, "Disruptive Challenges: Financial Implications and Strategic Responses to a Changing Retail Electric Business", Edison Electric Institute, January 2013. http://www.eei.org/ourissues/finance/documents/disruptivechallenges.pdf

[170] "How to Lose Half a Trillion Euros", The Economist, October 10, 2013. http://www. economist.com/news/briefing/21587782-europes-electricity-providers-face-existential-threat-how-lose-half-trillion-euros

[171] Baker, David, "Energy Storage Firm Tries No Cash Down Approach", San Francisco Chronicle, Oct 24, 2013, http://www.sfgate.com/business/article/Energy-storage-firm-tries-no-cash-down-approach-4922364.php

[172] Wikipedia contributors, "SunEdison, LLC", Wikipedia, the Free Encyclopedia, http://en.wikipedia.org/wiki/SunEdison_LLC

[173] "How much electricity does an American home use?", U.S. Energy Information Administration, http://www.eia.gov/tools/faqs/faq.cfm?id=97&t=3

[174] Baldwin Auck, Sara, "Utah Rising Storm for Net Metering", Solar Today, February 13, 2014, http://solartoday.org/2014/02/utah-rising-storm-for-net-metering

[175] Bellis, Mary, "History of the Digital Camera", About.com. http://inventors.about.com/library/inventors/bldigitalcamera.htm

[176] "NASA Hosts News Conference About 10 Years of Roving on Mars", Jet Propulsion Lab, California Institute of Technology, January 21, 2014 http://marsrovers.jpl.nasa.gov/newsroom/pressreleases/20140121a.html

[177] Changjan, Kenneth, "Mars Rover Marks an Unexpected Anniversary With a Mysterious Discovery", New York Times, January 23, 2014. http://www.nytimes.com/2014/01/24/science/space/mars-rover-marks-an-unexpected-anniversary-with-a-mysterious-discovery.html

[178] "What Is the Distance Between Earth and Mars?", Space.com, February 29, 2014. http://www.space.com/14729-spacekids-distance-earth-mars.html

[179] "Mars Artwork", Jet Propulsion Lab, California Institute of Technology. http://marsrovers.jpl.nasa.gov/gallery/artwork/rover3browse.html

[180] "The perils of extreme democracy", The Economist, April 20, 2011. http://www.economist.com/node/18586520

[181] "California Proposition 16, Supermajority Vote Required to Create a Community Choice Aggregator (June 2010)", Ballotpedia, http://ballotpedia.org/California_Proposition_16,_Supermajority_Vote_Required_to_Create_a_Community_Choice_Aggregator_(June_2010)

[182] Geesman, John, PG&E Ballot Initiative Factsheet, "Proposition 16: The Trojan Horse in California's June 8, 2010 Election", http://pgandeballotinitiativefactsheet.blogspot.com/

[183] Fahn, Larry, "Marin Voice: Prop. 16 is PG&E's power play", Marin Independent Journal, June 1, 2010. http://www.marinij.com/ci_15201680

[184] "California Proposition 16, Supermajority Vote Required to Create a Community Choice Aggregator (June 2010)", Ballotpedia. http://ballotpedia.org/California_Proposition_16,_Supermajority_Vote_Required_to_Create_a_Community_Choice_Aggregator_(June_2010)

[185] Pinnacle West Capital Corp Executive Compensation, Morningstar.com, http://insiders.morningstar.com/trading/executive-compensation.action?t=PNW&region=USA&culture=en-US

[186] "Mission", Arizona Corporation Commission, http://www.azcc.gov/divisions/Utilities/

[187] Baker, Brandon, "Arizona Imposes Unprecedented Fee on Solar Energy Users", EcoWatch, November 19, 2013, http://ecowatch.com/2013/11/19/arizona-imposes-unprecedented-fee-on-solar-energy-users/

[188] MacKenzie, Angus, "2013 Motor Trend Car of the Year: Tesla Model S", Motor Trend, January 2013, http://www.motortrend.com/oftheyear/car/1301_2013_motor_trend_car_of_the_year_tesla_model_s/viewall.html

[189] MacKenzie, Angus, "2013 Motor Trend Car of the Year: Tesla Model S", Motor Trend, January 2013, http://www.motortrend.com/oftheyear/car/1301_2013_motor_trend_car_of_the_year_tesla_model_s/viewall.html

[190] Oshman, Alan, "Elon Musk on Tesla Merger Prospects: Apple Has Got a Lot of Cash", Bloomberg, May 9, 2013, http://go.bloomberg.com/tech-deals/2013-05-09-elon-musk-on-tesla-merger-prospects-apple-has-a-lot-of-cash/

[191] Voelcker, John, "Does Tesla Already Outsell Audi, BMW, Lexus & Mercedes-Benz?", Green Car Reports, April 17, 2013: http://www.greencarreports.com/news/1083585_does-tesla-already-outsell-audi-bmw-lexus-mercedes-benz

[192] Wood, Column, "Global Vehicle Sales Hit Record 82 Million Units in 2012", AutoGuide.com, Mar 15, 2013, http://www.autoguide.com/auto-news/2013/03/global-vehicle-sales-hit-record-82-million-units-in-2012.html

[193] Ramsey, Mike, "Tesla's Stock is Outrunning Its Superfast Electric Car", Wall Street Journal, August 7, 2013. http://online.wsj.com/article/SB1000142412788732342060457865218036 0274840.html

[194] Ramsey, Mike, "Tesla's Stock is Outrunning Its Superfast Electric Car", Wall Street Journal, August 7, 2013. http://online.wsj.com/article/SB1000142412788732342060457865218036 0274840.html

[195] "Tesla Model S Achieves Best Safer Rating of Any Car Ever Tested", August 19, 2013, Tesla Motors website. http://www.teslamotors.com/about/press/releases/tesla-model-s-achieves-best-safety-rating-any-car-ever-tested

[196] Valdes-Dapena, Peter, "Tesla Model S Gets Consumer Reports Recommendation", CNNMoney, Oct 28, 2013, http://money.cnn.com/2013/10/28/autos/tesla-model-s-consumer-reports-recommended/

[197] Smith, Aaron, "GM names Mary Barra as CEO - first woman to run major automaker", CNN Money, December 10, 2013.
http://money.cnn.com/2013/12/10/news/companies/gm-ceo-mary-barra/

[198] "Fuel Source: Where the Energy Goes", Energy Requirements for Combined City/Highway Driving, U.S. Department of Energy, http://www.fueleconomy.gov/feg/atv.shtml

[199] Wikipedia contributors, "Engine Efficiency", Wikipedia, the Free Encyclopedia, http://en.wikipedia.org/wiki/Engine_efficiency

[200] Wikipedia contributors, "Energy Conversion Efficiency", Wikipedia, the Free Encyclopedia, http://en.wikipedia.org/wiki/Energy_conversion_efficiency

[201] "Using Energy Efficiently", Tesla Motors, http://www.teslamotors.com/goelectric/efficiency

[202] "What That Car Really Costs to Own", Consumer Reports, August 2012, http://www.consumerreports.org/cro/2012/12/what-that-car-really-costs-to-own/index.htm

[203] "Undergraduate Costs 2013-2014", Florida State University, retrieved Aug 3, 2103, http://admissions.fsu.edu/freshman/finances/costs.cfm

[204] "360 Degree Perspective of the North American Automotive Aftermarket", Frost & Sullivan, February 2011, retrieved August 3, 2013, http://www.slideshare.net/soaringvjr/north-american-auto-aftermarket-frost-0211

[205] "360 Degree Perspective of the North American Automotive Aftermarket", Frost & Sullivan, February 2011 retrieved August 3, 2013, http://www.slideshare.net/soaringvjr/north-american-auto-aftermarket-frost-0211

[206] "360 Degree Perspective of the North American Automotive Aftermarket", Frost & Sullivan, February 2011, retrieved August 3, 2013, http://www.slideshare.net/soaringvjr/north-american-auto-aftermarket-frost-0211

[207] Wikipedia contributors, "Wireless Power", Wikipedia, the Free Encyclopedia, http://en.wikipedia.org/wiki/Wireless_power

[208] Conductix-Wampfler website, available at http://www.conductix.us/en/markets/e-mobility

[209] Conductix-Wampfler website, available at http://www.conductix.us/en/markets/e-mobility

[210] Kharif, Olga, and Higgins, Tim, "GM to offer wireless charging for smartphones in some 2014 cars", Detroit Free Press, August 19, 2013, http://www.freep.com/article/20130819/BUSINESS0101/308190112/General-Motors-wireless-charging

[211] "Model X", Tesla Motors website, http://www.teslamotors.com/modelx

[212] "Commercial Version of the MIT Media Lab CityCar Unveiled at European Union Commission Headquarters", MIT Media Lab, http://www.media.mit.edu/news/citycar

[213] Dan Myggen's presentation at the Silicon Valley "Driving Charged and Connected" conference in Palo Alto, June 2013, http://svlg.org/policy-areas/transportation/charged-event-2013/silicon-valley-driving-charged-and-connected-agenda

[214] Kumparak, Greg, and Etherington, Darrell, "Live From Apple's iPhone 5S Announcement", TechCrunch, Sept 10, 2013. http://techcrunch.com/2013/09/10/live-blog-from-apples-iphone-5s-announcement/

[215] U.S. Department of Transportation, Bureau of Transportation Statistics, http://www.bts.gov/publications/national_transportation_statistics/html/table_01_32.html

[216] "How Big Box Stores like Wal-Mart Affect the Environment and Communities", Sierra Club, http://www.sierraclub.org/sprawl/reports/big_box.asp

[217] Langton, Adam, and Crisostomo, Noel, "Vehicle - Grid Integration -A Vision for Zero-Emission Transportation Interconnected throughout California's Electricity System", Emerging Procurement Strategies Section, Energy Division, California Public Utilities Commission, R. 13-11-XXX, October 2013

[218] "What Are P2P Communications", Skype company website, https://support.skype.com/en/faq/FA10983/what-are-p2p-communications

[219] Langton, Adam, and Crisostomo, Noel, "Vehicle - Grid Integration -A Vision for Zero-Emission Transportation Interconnected throughout California's Electricity System", Emerging Procurement Strategies Section, Energy Division, California Public Utilities Commission, R. 13-11-XXX, October 2013

[220] Herron, David, "GM: Next Generation Volt Will Be $10,000 Cheaper to Build", PlugInCars.com, May 1, 2013, http://www.plugincars.com/next-generation-volt-will-be-10k-cheaper-127121.html

[221] Duffer, Robert, "Tesla owner completes first coast-to-coast trip in electric vehicle via supercharging network", Chicago Tribune, January 27, 2014, http://articles.chicagotribune.com/2014-01-27/classified/chi-tesla-first-coast-to-coast-electric-vehicle_1_tesla-ceo-elon-musk-tesla-model-s-electric-vehicle

[222] Hull, Dana, "Coast to coast in a Tesla Model S, using only free Superchargers", San Jose Mercury News, January 28, 2014, http://www.mercurynews.com/business/ci_25010333/kentucky-man-drives-coast-coast-his-tesla-model

[223] "Supercharger", Tesla company website, http://www.teslamotors.com/supercharger

[224] "What That Car Really Costs to Own", Consumer Reports, August 2012, retrieved August 3, 2103, http://www.consumerreports.org/cro/2012/12/what-that-car-really-costs-to-own/index.htm

[225] "Using analytics to turbocharge performance in automotive marketing and incentive design", Accenture, Report ACC10-2623 / 11-2495, 2010, http://www.accenture.com/SiteCollectionDocuments/PDF/Accenture_Automotive_Sales_Analytics.pdf

[226] "Model S - Features & Specs", Tesla company website, http://www.teslamotors.com/models/features#/battery

[227] Healey, James R., "Report: Average price of new car hits record in August", USA Today, September 5, 2013. http://www.usatoday.com/story/money/cars/2013/09/04/record-price-new-car-august/2761341/

[228] "General Motors – Key Ratios", Morningstar.com, http://financials.morningstar.com/ratios/r.html?t=GM

[229] "Bayerische Motoren Werke AG – Key Ratios", Morningstar.com, http://financials.morningstar.com/ratios/r.html?t=XFRA:BMW

[230] Gordon-Bloomfield, Nikki, "U.S. Sec. Of Energy: Cheaper Batteries Mean More Electric Cars", Green Car Reports, January 11, 2012, http://www.greencarreports.com/news/1071597_u-s-sec-of-energy-cheaper-batteries-mean-more-electric-cars

[231] Meggison, Andrew, "The Changing Cost of Electric Cars", Gas2.org http://gas2.org/2013/06/13/the-changing-price-of-electric-cars/

[232] Tesla Gigafactory presentation, http://www.teslamotors.com/sites/default/files/blog_attachments/gigafactory.pdf

[233] "Panasonic, Tesla to set up auto battery plant in US", Nikkei Asian Review, February 26, 2014. http://asia.nikkei.com/Business/Deals/Panasonic-Tesla-to-set-up-auto-battery-plant-in-US

[234] Vance, Ashlee, "Tesla's Industrial-Grade Solar Power Storage System", Bloomberg BusinessWeek, December 6, 2013. http://www.businessweek.com/articles/2013-12-06/teslas-solar-power-storage-unit

[235] Galveset al, "Tesla Motors: Only Just Begun, Upgrading to Buy",, Deutsche Bank Markets Research, July 26, 2013

[236] "Liquid Metal Batteries", GroupSadoway, Massachusetts Institute of Technology. http://sadoway.mit.edu/research/liquid-metal-batteries

[237] LaMonica, Martin, "Ambri's Better Battery", MIT Technology Review, February 18, 2013, http://www.technologyreview.com/featuredstory/511081/ambris-better-grid-battery/

[238] Yi Cui, "Energy Seminar", Stanford Precourt Institute for Energy, February 3, 2013, http://www.youtube.com/watch?v=OZ7cEWrX9U4

[239] "Model S – Order", Tesla Motors website, http://www.teslamotors.com/models/design

[240] Galves, Danet al, "Tesla Motors: Only Just Begun, Upgrading to Buy", Deutsche Bank Markets Research, July 26, 2013

[241] "SUV Rankings – Affordable Midsize SUVs", US News & World Report, http://usnews.rankingsandreviews.com/cars-trucks/rankings/Affordable-Midsize-SUVs/

[242] Undercoffer, David, "Tesla Motors plans to debut cheaper car in early 2015", Los Angeles Times, December 15, 2013 http://articles.latimes.com/2013/dec/15/autos/la-fi-hy-autos-tesla-model-e-debut-2015-20131213

[243] Cameron, Kevin, "Questions Linger on Battery Prices in Electric Cars", New York Times, October 23, 2012, http://www.nytimes.com/2012/10/24/business/energy-environment/questions-linger-on-battery-prices-in-electric-car-industry.html

[244] Hensley, Russelet al, "Battery Technology Charges Ahead", McKinsey Quarterly, July 2012

[245] "Tesla Motors", MIT Technology Review, Vol 117, No 2, March/April, 2014

[246] Healey, James R., "Report: Average price of new car hits record in August", USA Today, September 5, 2013. http://www.usatoday.com/story/money/cars/2013/09/04/record-price-new-car-august/2761341/

[247] "Zipcar at a Glance", Company Overview, Fall 2012

[248] Clothier, Mark, "Zipcar Soars After Profit Topped Analysts' Estimates", Bloomberg, Nov 9, 2012. http://www.bloomberg.com/news/2012-11-09/zipcar-soars-after-profit-topped-analysts-estimates.html

[249] Duerson, Meena Hart, "Airbnb founder wants you to open your doors to strangers — and let them sleep over", Today, March 29, 2013, http://www.today.com/news/airbnb-founder-wants-you-open-your-doors-strangers-let-them-1C9138916

[250] Airbnb.com company website. Retrieved November 12, 2103: https://www.airbnb.com/about/about-us

[251] CouchSurfing.com company website. Retrieved July 30, 2103: http://blog.couchsurfing.com/

[252] Panzarino, Matthew, "Leaked Uber Numbers, Which We've Confirmed, Point To Over $1B Gross, $213M Revenue", TechCrunch, December 4, 2013. http://techcrunch.com/2013/12/04/leaked-uber-numbers-which-weve-confirmed-point-to-over-1b-gross-revenue-213m-revenue/

[253] Swisher, Kara, "Uber Filing in Delaware Shows TPG Investment at $3.5 Billion Valuation; Google Ventures Also In". AllThingsD, August 22, 2013. http://allthingsd.com/20130822/uber-filing-in-delaware-shows-tpg-investment-at-3-5-billion-valuation-google-ventures-also-in/

[254] Sousanis, John, "World Vehicle Population Tops 1 Billion Units", WardsAuto, August 15, 2011 http://wardsauto.com/ar/world_vehicle_population_110815

[255] "Transport Outlook 2011 - Meeting the needs of 9 Billion People", International Transport Forum, http://www.internationaltransportforum.org/Pub/pdf/11Outlook.pdf

[256] Wikimedia Commons, "Google Lexus RX 450h Self-Driving Car", http://upload.wikimedia.org/wikipedia/commons/1/1b/Google%27s_Lexus_RX_450h_Self-Driving_Car.jpg

[257] Silberg, Gary, and Wallace, Richard, "Self-Driving Cars: The Next Revolution", KPMG

[258] Wikipedia contributors, "United States Military Casualties of War", Wikipedia, the Free Encyclopedia, http://en.wikipedia.org/wiki/United_States_military_casualties_of_war

[259] Motor Vehicle Traffic Fatalities & Fatality Rate: 1899-2003 (Based on Historical NHTSA and FHWA Data", Advocates for Highway and Auto Safety, http://www.saferoads.org/federal/2004/TrafficFatalities1899-2003.pdf

[260] "Global Status Report on Road Safety 2013 – Time for Action", World Health Organization, http://www.who.int/violence_injury_prevention/road_safety_status/2013/report/en/

[261] "Global Status Report on Road Safety - Time for Action", World Health Organization,

[262] "Global Status Report on Road Safety 2013 – Time for Action", World Health Organization

[263] Mullainathan, Sendil, "Get Some Sleep, and Wake Up the G.D.P.", The New York Times, February 2, 2014, http://www.nytimes.com/2014/02/02/business/get-some-sleep-and-wake-up-the-gdp.html

[264] Gross, Bill, "Google's Self Driving Car Gathers Nearly 1 GB/Sec", LinkedIn Today, May 2, 2013, http://www.linkedin.com/today/post/article/20130502024505-9947747-google-s-self-driving-car-gathers-nearly-1-gb-per-second

[265] Rodrigue, Jean-Paul , "The Geography of Transport Systems", third edition, New York: Routledge, 2013, http://people.hofstra.edu/geotrans/eng/ch6en/conc6en/ch6c1en.html

[266] Litman, Todd, "Smart Congestion Relief - Comprehensive Evaluation Of Traffic Congestion Costs and Congestion Reduction Strategies", Victoria Transport Policy Institute, January 29, 2014

[267] Shladover, Steven E., "Highway Capacity Increases From Automated Driving", California PATH Program, 25 July 2013 presentation, available at http://onlinepubs.trb.org/onlinepubs/conferences/2012/Automation/presentations/Shladover2.pdf

[268] Shladover, Steven E., "Highway Capacity Increases From Automated Driving", California PATH Program, 25 July 2013 presentation, available at http://onlinepubs.trb.org/onlinepubs/conferences/2012/Automation/presentations/Shladover2.pdf

[269] Ackerman, Evan, "Study: Intelligent Cars Could Boost Highway Capacity by 273%", IEEE Spectrum, Sept 4, 2012, http://spectrum.ieee.org/automaton/robotics/artificial-intelligence/intelligent-cars-could-boost-highway-capacity-by-273

[270] Schrank, Davidet al, "TTI's 2012 URBAN MOBILITY REPORT", Texas A&M Transportation Institute, December 2012, http://mobility.tamu.edu,

[271] Silberg, Gary, and Wallace, Richard,"Self-Driving Cars: The Next Revolution", KPMG

[272] Jaffe, Eric, "Has the Rise of Online Shopping Made Traffic Worse?", Atlantic Cities, August 2, 2013. http://www.theatlanticcities.com/commute/2013/08/has-rise-online-shopping-made-traffic-worse/6409/

[273] Jaffe, Eric, "Has the Rise of Online Shopping Made Traffic Worse?", Atlantic Cities, August 2, 2013. http://www.theatlanticcities.com/commute/2013/08/has-rise-online-shopping-made-traffic-worse/6409/

[274] "Road Frustration Index", Senseable City Lab, http://storify.com/SenseableCity/road-frustration-index

[275] Mitchell, William, "Personal Mobility", http://h20.media.mit.edu/pdfs/wjm2007-0509.pdf

[276] Silberg, Gary, and Wallace, Richard, "Self-Driving Cars: The Next Revolution", KPMG

[277] Ayres, Tom, "How Self-Driving Cars Could Turn Our Roads Green", EnterpriseTech, May 28, 2013. http://www.digitalmanufacturingreport.com/dmr/2013-05-28/how_self-driving_cars_could_turn_our_roads_green.html

[278] "Your Car Costs – How Much Are you really Paying to Drive", AAA, 2012 edition, at http://newsroom.aaa.com/wp-content/uploads/2012/04/YourDrivingCosts2012.pdf

[279] Berman, Jillian, "U.S. Median Annual Wage Falls To $26,364 As Pessimism Reaches 10-Year High", The Huffington Post, January 23, 2012. http://www.huffingtonpost.com/2011/10/20/us-incomes-falling-as-optimism-reaches-10-year-low_n_1022118.html

[280] "Global Report on Road Safety - Time for Action", World Health Organization,

[281] Silberg, Gary, and Wallace, Richard, "Self-Driving Cars: The Next Revolution", KPMG

[282] Litman, Todd, "Smart Congestion Relief - Comprehensive Evaluation Of Traffic Congestion Costs and Congestion Reduction Strategies", Victoria Transport Policy Institute, January 29, 2014

[283] Foy, Henry, and Bryant, Chris"Nissan Promises Self-Driving Cars by 2020", Financial Times, August 27, 2013. http://www.ft.com/cms/s/0/b8fad15e-0f3c-11e3-ae66-00144feabdc0.html

[284] Nissan Newsroom, "Nissan Announces Self-Driving Car by 2020", August 27, 2013, http://www.youtube.com/watch?v=NWd2Eoxfcvw

[285] "Automotive Industry – Global Data", Cisco Customer Experience Research, Cisco Systems, May 2013

[286] "Preliminary Statement of Policy Concerning Automated Vehicles", National Highway Traffic Safety Administration

[287] "2012 Audi Driver Assistance Systems", Audi company website, http://audiusanews.com/imagegallery/view/215/47/2012-audi-a6-driver-assistance-systems

[288] Mercedes Benz BAS Plus with Cross Traffic Assist, company website, http://techcenter.mercedes-benz.com/en/bas_plus_cross_traffic_assist/detail.html

[289] http://www.consumerreports.org/cro/news/2013/06/bmw-traffic-jam-assistant-puts-self-driving-car-closer-than-you-think/index.htm

[290] Sladover, Steven E., "Why Automated Vehicles Need to Be Connected Vehicles", University of California PATH Program, IEEE Vehicular Networking Conference December 17, 2013

[291] "BMW Traffic Jam Assistant puts self-driving car closer than you think", Consumer Reports News: June 11, 2013, http://content.usatoday.com/communities/driveon/post/2012/06/google-discloses-costs-of-its-driverless-car-tests/1

[292] Wikipedia contributors, "Lidar", Wikipedia, the Free Encyclopedia, http://en.wikipedia.org/wiki/Lidar

[293] Trinder, John, "Current Trends in Photogrammetry and Imaging including Lidar", University of New South Wales,

[294] Bilger, Burkhard, "Auto Correct – Has the Self-Driving Car At Last Arrived?", The New Yorker, Nov 25, 2013, http://www.newyorker.com/reporting/2013/11/25/131125fa_fact_bilger

[295] "Fujitsu Semiconductor Develops World's First 360° Wraparound View System with Approaching Object Detection", company press release, May 16, 2013, http://jp.fujitsu.com/group/fsl/en/release/20130516.html

[296] "Fujitsu Semiconductor Develops World's First 360° Wraparound View System with Approaching Object Detection", company press release, May 16, 2013, http://jp.fujitsu.com/group/fsl/en/release/20130516.html

[297] "Ford Developer Program", Ford company website https://developer.ford.com/

[298] Wikipedia contributors, "History of the iPhone", Wikipedia, Wikipedia, the Free Encyclopedia, http://en.wikipedia.org/wiki/History_of_the_iPhone

[299] Wikipedia contributors, "Android Version History", Wikipedia,,the Free Encyclopedia, http://en.wikipedia.org/wiki/Android_version_history

[300] Bradley, Tony, "Android Dominates Market Share, But Apple Makes All The Money", November 15, 2013, http://www.forbes.com/sites/tonybradley/2013/11/15/android-dominates-market-share-but-apple-makes-all-the-money/

[301] Carr, Austin, "J.Crew CEO, Apple Board Member Mickey Drexler Reveals Steve Jobs' iCar Dream, Confirms "Living Room" Plans", FastCompany, May 16, 2012, http://www.fastcompany.com/1837636/j-crew-ceo-apple-board-member-mickey-drexler-reveals-steve-jobs-icar-dream-confirms-living-r

[302] Clothier, Mark, "Zipcar Soars After Profit Topped Analysts' Estimates", Bloomberg, November 9, 2012, http://www.bloomberg.com/news/2012-11-09/zipcar-soars-after-profit-topped-analysts-estimates.html

[303] Wood, Column, "Global Vehicle Sales Hit Record 82 Million Units in 2012", AutoGuide.com, March 15, 2013, http://www.autoguide.com/auto-news/2013/03/global-vehicle-sales-hit-record-82-million-units-in-2012.html

[304] Upbin, Bruce, "Monsanto Buys Climate Corp for $930 million", Forbes.com, October 2, 2013

[305] Gross, Bill, "Google's Self Driving Car Gathers Nearly 1 GB/Sec", LinkedIn Today, May 2, 2013

[306] Wikipedia contributors, "Chernobyl Disaster", Wikipedia,,the Free Encyclopedia, http://en.wikipedia.org/wiki/Chernobyl_disaster

[307] Gorbachev, Michael, "Turning Point at Chernobyl", Project Syndicate, April 14, 2006, http://www.project-syndicate.org/commentary/gorbachev3/English

[308] Morin, Hervé, "L'effet de Tchernobyl en France a été jusqu'à mille fois sous-évalué", Le Monde, April 24, 2006, http://www.lemonde.fr/planete/article/2006/04/24/l-effet-de-tchernobyl-en-france-a-ete-jusqu-a-mille-fois-sous-evalue_764692_3244.html

[309] Morin, Hervé, "L'effet de Tchernobyl en France a été jusqu'à mille fois sous-évalué", Le Monde, April 24, 2006

[310] Cherry, Steven, "Crowdsourcing Radiation Monitoring", IEEE Spectrum, November 17, 2011. http://spectrum.ieee.org/podcast/geek-life/hands-on/crowdsourcing-radiation-monitoring

[311] "The bGeigie Nano Geiger Counter Kit", International Medcom website, http://medcom.com/radiation-monitors/geiger-counters/bgeigie-kit/ Accessed Aug 19, 2013

[312] "The bGeigie Nano Geiger Counter Kit", International Medcom website, http://medcom.com/radiation-monitors/geiger-counters/bgeigie-kit/

[313] Wikipedia contributors, "Regulatory Capture", Wikipedia „the Free Encyclopedia http://en.wikipedia.org/wiki/Regulatory_capture

[314] McKenna, John, "Sellafield clean-up costs out of control", Process Engineering, Feb 3, 2013, http://processengineering.theengineer.co.uk/sellafield-clean-up-costs-out-of-control/1015427.article

[315] Mulkern, Anne C., "Nuclear Energy: Who'll get stuck with San Onofre's $3B tab?", E&E News, June 10, 2013, http://www.eenews.net/stories/1059982573

[316] Martin, James, "Solar PV Price Index–July 2013", Solar Choice, July 4, 2013.

[317] Wikipedia contributors, "Energy in the United Kingdom", Wikipedia„the Free Encyclopedia, http://en.wikipedia.org/wiki/Energy_in_the_United_Kingdom

[318] Jowit, Juliette, "Nuclear power: ministers offer reactor deal until 2050", The Guardian, February, 18, 2013, http://www.theguardian.com/environment/2013/feb/18/nuclear-power-ministers-reactor

[319] Jowit, Juliette, "Nuclear power: ministers offer reactor deal until 2050", The Guardian, February, 18, 2013, http://www.theguardian.com/environment/2013/feb/18/nuclear-power-ministers-reactor

[320] Meikle Capital, Technology Equilibrium Fund, LP, July 8, 2013 newsletter

[321] Wikipedia contributors, "Sellafield", Wikipedia, the Free Encyclopedia, http://en.wikipedia.org/wiki/Sellafield

[322] "Sellafield clean-up costs out of control", Process Engineering, Feb 3, 2013, http://processengineering.theengineer.co.uk/sellafield-clean-up-costs-out-of-control/1015427.article

[323] Wikipedia contributors, "Vogtle Electric Generating Plant", Wikipedia„the Free Encyclopedia, http://en.wikipedia.org/wiki/Vogtle_Electric_Generating_Plant

[324] "Learning Curve Calculator", Federation of American Scientists, http://www.fas.org/news/reference/calc/learn.htm

[325] "Learning Curve Calculator", Federation of American Scientists, http://www.fas.org/news/reference/calc/learn.htm

[326] Koomey, Jonathan, and Holtman, Nathan, "A reactor-level analysis of busbar costs for US nuclear plants, 1970–2005"„ Energy Policy, 2007

[327] Harding, Jim, "Economics of New Nuclear Power and Proliferation Risks in a Carbon-Constrained World", Nonproliferation Policy Education Center, 2007

[328] "Nuclear Safety and Nuclear Economics", Mark Cooper, Ph. D., Senior Fellow for Economic Analysis Institute for Energy and the Environment Vermont Law School, Symposium on the Future of Nuclear Power University of Pittsburgh March 27-28, 2012

[329] "Nuclear Safety and Nuclear Economics", Mark Cooper, Ph. D., Senior Fellow for Economic Analysis Institute for Energy and the Environment Vermont Law School, Symposium on the Future of Nuclear Power University of Pittsburgh March 27-28, 2012

[330] "Myths and Facts about Economics and Financing", Nuclear Energy Institute, http://www.nei.org/Knowledge-Center/Backgrounders/Fact-Sheets/Myths-Facts-About-Economics-Financing

[331] Wikipedia contributors, "Energy Policy Act of 2005", Wikipedia,,the Free Encyclopedia, http://en.wikipedia.org/wiki/Energy_Policy_Act_of_2005

[332] Sturgis, Sue, "Power Politics: Big Nuclear Money Grab", Institute for Southern Studies, March 2, 2009, http://www.southernstudies.org/2009/03/power-politics-big-nuclears-money-grab.html

[333] "Obama Administration Announces Loan Guarantees to Construct New Nuclear Power Reactors in Georgia", White House press release, Feb 16, 2010, http://www.whitehouse.gov/the-press-office/obama-administration-announces-loan-guarantees-construct-new-nuclear-power-reactors

[334] "DOE Loan Guarantee Program: Vogtle Reactors 3&4", Taxpayers For Common Sense, February 19, 2014 http://www.taxpayer.net/library/article/doe-loan-guarantee-program-vogtle-reactors-34

[335] Smith, Rebecca, "New Wave of Nuclear Plants Faces High Costs," The Wall Street Journal, May 12, 2008, http://online.wsj.com/article/SB121055252677483933.html?mod=hpp_us_whats_news#articleTabs%3Darticle

[336] Gore, Al, Our Choice, Bloomsbury, 2009, p. 157.

[337] Lovins, Amory, Sheikh, Imram, and Markevich, Alex. "Nuclear Power: Climate Fix or Folly", 2009

[338] Smith, Rebecca, "New Wave of Nuclear Plants Faces High Costs," The Wall Street Journal, May 12, 2008

[339] Wikipedia contributors, "Washington Public Power Supply System", Wikipedia,,the Free Encyclopedia, http://en.wikipedia.org/wiki/Washington_Public_Power_Supply_System

[340] Wikipedia contributors, "Columbia Generating Station", Wikipedia,,the Free Encyclopedia, http://en.wikipedia.org/wiki/Columbia_Generating_Station

[341] Lomax, Simon, and Snyder, Jim, "Obama would Triple Guarantees for Building Nuclear Reactors", Bloomberg, Feb 14, 2011, http://www.bloomberg.com/news/2011-02-14/obama-would-triple-guarantees-for-building-nuclear-reactors.html

[342] Pawlawski, A, "Aviation safety rate: One accident for every 1.4 million flights", CNN Travel, Feb 22, 2010, http://www.cnn.com/2010/TRAVEL/02/22/aviation.safety.report/

[343] Mark A. Ruffalo, Marco Krapels and Mark Z. Jacobson: "Power the World with Wind, Water and Sunlight", June 20, 2012, http://www.youtube.com/watch?v=N_sLt5gNAQs

[344] Sieg, Linda, and Takenaka, Kiyoshi, "Japan secrecy act stirs fears about press freedom, right to know", Reuters, October 24, 2013, http://www.reuters.com/article/2013/10/24/us-japan-secrecy-idUSBRE99N1EC20131024

[345] "Nuclear Power in the World Today", World Nuclear Association, April, 2014, http://www.world-nuclear.org/info/Current-and-Future-Generation/Nuclear-Power-in-the-World-Today/

[346] Wikipedia contributors, "Nuclear Power in France", Wikipedia, the Free Encyclopedia, http://en.wikipedia.org/wiki/Nuclear_power_in_France

[347] Wikipedia contributors, "Nuclear Power in Japan", Wikipedia, the Free Encyclopedia, http://en.wikipedia.org/wiki/Nuclear_power_in_Japan

[348] Wikipedia contributors, "Nuclear Power in the United States", Wikipedia, the Free Encyclopedia, http://en.wikipedia.org/wiki/Nuclear_power_in_the_United_States

[349] Günther, Benjaminet al, "Calculating a risk-appropriate insurance premium to cover third-party liability risks that result from operation of nuclear power plants", German Renewable Energy Federation (BEE), April 1, 2011, http://www.laka.org/docu/boeken/pdf/6-01-0-30-34.pdf

[350] "Brannon Solar Renewable Energy Contract", City Council Staff Report, City of Palo Alto, November 5, 2012.

[351] Wikipedia contributors, "List of Countries by GDP", Wikipedia, the Free Encyclopedia, http://en.wikipedia.org/wiki/List_of_countries_by_GDP_%28nominal%29

[352] "Russia GDP", Trading Economics, http://www.tradingeconomics.com/russia/gdp

[353] Gorbachev, Michael, "Turning Point at Chernobyl", Project Syndicate, April 14, 2006, http://www.project-syndicate.org/commentary/gorbachev3/English

[354] Kennedy, Duncan, "Italy nuclear: Berlusconi accepts referendum blow", BBC News, June 14, 2011, http://www.bbc.co.uk/news/world-europe-13741105

[355] "Europe to Decommission Majority of Nuclear Power Stations by 2030 While US Bucks Global Trend", GlobaData, June 6, 2012. http://energy.globaldata.com/media-center/press-releases/power-and-resources/europe-to-decommission-majority-of-nuclear-power-stations-by-2030-while-us-bucks-global-trend

[356] "60,000 protest Japan's plan to restart nuclear power plants", UPI, June 2, 2013. http://www.upi.com/Top_News/World-News/2013/06/02/60000-protest-Japans-plan-to-restart-nuclear-power-plants/UPI-34961370197818/

[357] Romm, Joe, "NRG to abandon two new South Texas nuclear plants, write down $481 million investment", ClimateProgress, April 26, 2011

[358] Douglass, Elizabeth, "First U.S. Nuclear Power Closures in 15 Years Signal Wider Problems for Industry", Sept, 24 2013. http://insideclimatenews.org/news/20130924/first-us-nuclear-power-closures-15-years-signal-wider-problems-industry

[359] Eggers, Dan et al, "Nuclear... The Middle Age Dilemma? - Facing Declining Performance, Higher Costs, Inevitable Mortality", CreditSuisse, February 19, 2013

[360] Cooper, Mark, "Renaissance in Reverse: Competition Pushes Aging U.S. Nuclear Reactors to the Brink of Economic Abandonment", Vermont Law School, July 18, 2013

[361] Douglass, Elizabeth, "First U.S. Nuclear Power Closures in 15 Years Signal Wider Problems for Industry", Sept, 24 2013.

[362] Eggers, Dan et al, "Nuclear... The Middle Age Dilemma? - Facing Declining Performance, Higher Costs, Inevitable Mortality", CreditSuisse, February 19, 2013

[363] Eggers, Dan et al, "Nuclear... The Middle Age Dilemma? - Facing Declining Performance, Higher Costs, Inevitable Mortality", CreditSuisse, February 19, 2013

[364] Patel, Tara, "EDF Writing 'Last Chapter' on Nuclear in U.S., Piquemal Says", Bloomberg, July 30, 2013 http://www.bloomberg.com/news/2013-07-30/edf-writing-last-chapter-on-nuclear-in-u-s-piquemal-says.html

[365] Wald, Matthew, "USEC, Enricher of Uranium for U.S., Seeks Bankruptcy", New York Times, December 16, 2013. http://www.nytimes.com/2013/12/17/business/energy-environment/usec-enricher-of-uranium-for-us-seeks-bankruptcy.html

[366] "USEC Portsmouth 'American Centrifuge Plant' project, USA", Wise Uranium. http://www.wise-uranium.org/epusecc.html

[367] Romm, Joe, "Exclusive analysis, Part 1: The staggering cost of new nuclear power", ClimateProgress, January 5, 2009. http://thinkprogress.org/climate/2009/01/05/202859/study-cost-risks-new-nuclear-power-plants/

[368] "Average Retail Price of Electricity to Ultimate Customers by End-Use Sector", U.S. Energy Information Administration. http://www.eia.gov/electricity/monthly/epm_table_grapher. cfm?t=epmt_5_6_a

[369] Gossens, Ehren, and Marin, Christopher, "First Solar: May Sell Solar at Less than Coal", Bloomberg News, February 1, 2013. http://www.bloomberg.com/news/2013-02-01/first-solar-may-sell-cheapest-solar-power-less-than-coal.html

[370] "Exelon scraps Texas reactor project", Nuclear Engineering International, August 29, 2012. http://www.neimagazine.com/news/newsexelon-scraps-texas-reactor-project-721

[371] Rago, Joseph, "A Life in Energy and (Therefore) Politics", Wall Street Journal, October 22, 2011. http://online.wsj.com/news/articles/SB10001424052970204618704576641351747987560

[372] Macalister, Terry, "Hinkley Point C nuclear subsidy plan queried by European commission", The Guardian, December 18, 2013. http://www.theguardian.com/business/2013/dec/18/hinkley-point-c-nuclear-subsidy-european-commission

[373] "Ivanpah Solar Electric Generating System Reaches 'First Sync' Milestone", Brightsource company press press release, September 24, 2013. http://www.brightsourceenergy.com/first-sync

[374] Mahdi, Wael, and Roca, Marc, "Saudi Arabia Plans $109 Billion Boost for Solar Power", Bloomberg, May 11, 2012. http://www.bloomberg.com/news/2012-05-10/saudi-arabia-plans-109-billion-boost-for-solar-power.html

[375] Mahdi, Wael, and Roca, Marc, "Saudi Arabia Plans $109 Billion Boost for Solar Power", Bloomberg, May 11, 2012

[376] "U.S. Crude Oil First Purchase Price", U.S. Energy Information Administration. http://www.eia.gov/dnav/pet/hist/LeafHandler.ashx?n=PET&s=F000000__3&f=A

[377] "1970 Economy / Prices", 1970sFlashback.com. http://www.1970sflashback.com/1970/economy.asp

[378] "BP Statistical Review of World Energy", June 2013, http://www.bp.com/statisticalreview

379 Sreekumar, Ariun, "Why Canada's Oil Sands Boom Could Turn to Bust", The Motley Fool, May 1, 2013. http://www.fool.com/investing/general/2013/05/01/why-canadas-oil-sands-boom-could-turn-to-bust.aspx

380 Sreekumar, Ariun, "Why Canada's Oil Sands Boom Could Turn to Bust", The Motley Fool, May 1, 2013

381 "Performance Profile of Major Energy Producers - 2009", Feb 2011, U.S. Energy Information Administration, http://www.eia.gov/finance/performanceprofiles/pdf/020609.pdf

382 "Petroleum and Other Liquids, Spot Prices", U.S. Energy Information Administration, downloaded July 19th, 2013: http://www.eia.gov/dnav/pet/pet_pri_spt_s1_a.htm

383 "Petroleum and Other Liquids, Spot Prices", U.S. Energy Information Administration,

384 Suo, Jenny, "Tokelau to become world's first solar-powered country", 3News, July 25, 2012 http://www.3news.co.nz/Tokelau-to-become-worlds-first-solar-powered-country/tabid/1160/articleID/262649/Default.aspx#ixzz22EO3tNx3

385 Wikipedia contributors, "Tokelau", Wikipedia, the Free Encyclopedia, http://en.wikipedia.org/wiki/Tokelau

386 "Tokelau 100% Powered', PowerSmart Solar company website, http://powersmartsolar.co.nz/our_projects/id/185/TOKELAU%20-%20100%25%20SOLAR%20POWERED

387 Wikipedia contributors, "Communications In India", Wikipedia, the Free Encyclopedia, http://en.wikipedia.org/wiki/Communications_in_India

388 Seba, Tony, "India Needs to Leapfrog to Solar and Electricity 2.0", Aug 8, 2012, http://tonyseba.com/cleanenergyeconomy/india-needs-to-leapfrog-to-solar-and-electricity-2-0/

389 "Developing Countries Subsidize Fossil Fuels, Artificially Lowering Prices", Institute for Energy Research, January 3, 2013. http://www.instituteforenergyresearch.org/2013/01/03/developing-countries-subsidize-fossil-fuel-consumption-creating-artificially-lower-prices/

390 André-Jacques Auberton-Hervé presentation at the Intersolar NorthAmerica CEO panel, July 9th, 2013.

391 "Energy Subsidy Reform: Lessons and Implications", International Monetary Fund, January 28, 2013

392 Wang, Yue, "More People Have Cell Phones Than Toilets, U.N. Study Shows", Time, March 25, 2013.

[393] DeGunther, Rik, "How Large Does Your Solar Power System Need to Be?", from Solar Power Your Home For Dummies, 2nd Edition, http://www.dummies.com/how-to/content/how-large-does-your-solar-power-system-need-to-be.seriesId-246925.html

[394] Narasimha Rao et al, "An overview of Indian Energy Trends: Low Carbon Growth and Development Challenges", Prayas Energy Group, India, September 2009
[395] "Green stoves to replace challahs", Times of India, December 3, 2009, http://timesofindia.indiatimes.com/india/Green-stoves-to-replace-chullahs/articleshow/5293563.cms

[396] U.S. Department of Transportation, Bureau of Transportation Statistics, http://www.bts.gov/publications/national_transportation_statistics/html/table_01_32.html

[397] "All-Electric Vehicles: Compare Side-by-Side", FuelEconomy.gov, U.S. Department of Energy, http://www.fueleconomy.gov/feg/evsbs.shtml

[398] Berman, Brad, "What It Takes to Get 100 Miles of Range in My Electric Car", PlugInCars.com, August 25, 2011, http://www.plugincars.com/what-it-takes-get-100-miles-range-my-electric-car-107677.html

[399] "All-Electric Vehicles: Compare Side-by-Side", U.S. Department of Energy http://www.fueleconomy.gov/feg/evsbs.shtml

[400] Willis, Ben, "Amonix beats own record with 35.9% CPV module efficiency", PV Tech, August 21, 2013, http://www.pv-tech.org/news/amonix_beats_own_record_with_35.9_cpv_module_efficiency

[401] "The Truth About America's Energy: Big Oil Stockpiles Supplies and Pockets Profits," A Special Report by the Committee on Natural Resources Majority Staff," U.S. House of Representatives, Committee on Natural Resources, Rep Nick J. Hall – Chairman, June 2008

[402] "How Big Box Stores like Wal-Mart Affect the Environment and Communities", Sierra Club, http://www.sierraclub.org/sprawl/reports/big_box.asp

[403] Wikipedia contributors, "2010 San Bruno Explosion", Wikipedia, the Free Encyclopedia, retrieved July 21, 2013, http://en.wikipedia.org/wiki/2010_San_Bruno_pipeline_explosion

[404] Photo: Brocken Inaglory, Source: Wikimedia, retreived July 21, 2013, https://en.wikipedia.org/wiki/File:Devastation_in_San_Bruno.jpg

[405] Wikipedia contributors, "The 1906 San Francisco Earthquake", Wikipedia, the Free Encyclopedia http://en.wikipedia.org/wiki/1906_San_Francisco_earthquake

[406] Alleman, James E., and Moseman, Brooke, "Asbestos Revisited", Scientific American, July 1997

[407] Wikipedia contributors, "Greenhouse Gas", Wikipedia, the Free Encyclopedia, http://en.wikipedia.org/wiki/Greenhouse_gas

[408] "Natural Gas Transmission Leakage Rates", SourceWatch Center for Media and Democracy, retrieved July 21, 2013, http://www.sourcewatch.org/index.php?title=Natural_gas_transmission_leakage_rates

[409] Revkin, Andrew, and Krauss, Clifford, "Curbing Emissions by Sealing Gas Leaks", New York Times, Oct 14, 2009, http://www.nytimes.com/2009/10/15/business/energy-environment/15degrees.html

[410] Wikipedia contributors, "2010 San Bruno Pipeline Explosion", Wikipedia, the Free Encyclopedia, http://en.wikipedia.org/wiki/2010_San_Bruno_pipeline_explosion

[411] "About Natural Gas Pipelines", Energy Information Administration, http://www.eia.gov/pub/oil_gas/natural_gas/analysis_publications/ngpipeline/index.html

[412] "The State of the National Pipeline Infrastructure", U.S. Department of Transportation,

[413] Revkin, Andrew, and Krauss, Clifford, "Curbing Emissions by Sealing Gas Leaks", New York Times, Oct 14, 2009

[414] Mann, Charles, "What if we Never Run out of Oil", The Atlantic, April 24, 2013, http://www.theatlantic.com/magazine/archive/2013/05/what-if-we-never-run-out-of-oil/309294/5/

[415] Lucas, Tim, "The Dangers Underneath", Duke Today, January 16, 2014, http://today.duke.edu/2014/01/dcgas

[416] Wikipedia contributors, "Greenhouse Gas", Wikipedia, the Free Encyclopedia, http://en.wikipedia.org/wiki/Greenhouse_gas

[417] "We're Working to do the right thing, every day", PG&E advertising page A3, San Francisco Chronicle, October 31, 2013

[418] PG&E Corporation and Pacific Gas and Electric 2012 Annual Report", http://investor.pgecorp.com/files/doc_downloads/2012_Annual_Report.pdf

[419] "The Next Shock?", The Economist, March 4th, 1999, http://www.economist.com/node/188181

[420] "The Next Shock?", The Economist, March 4th, 1999,

[421] FAQ: How many gallons of gasoline does one barrel of oil make?", U.S. Energy Information Administration, http://www.eia.gov/tools/faqs/faq.cfm?id=24&t=10

[422] Mouawad, Jad, "One Year After Oil's Price Peak: Volatility", The New York Times, July 10, 2009, http://green.blogs.nytimes.com/2009/07/10/one-year-after-oils-price-peak-volatility/

[423] "U.S. Crude Oil First Purchase Price", U.S. Energy Information Administration, http://www.eia.gov/dnav/pet/hist/LeafHandler.ashx?n=pet&s=f000000__3&f=m

[424] "U.S. Natural Gas Wellhead Price", U.S. Energy Information Administration, http://www.eia.gov/dnav/ng/hist/n9190us3A.htm

[425] "U.S. Price of Natural Gas Delivered to Residential Customers", U.S. Energy Information Administration, http://www.eia.gov/dnav/ng/hist/n3010us3M.htm

[426] International Energy Agency, "World Energy Outlook 2012 – Executive Summary", www.worldenergyoutlook.org

[427] "The Next Qatar?", The Economist, July 27, 2013, retrieved August 5, 2013, http://www.economist.com/news/business/21582272-cost-exploiting-australias-new-found-gas-supplies-soaring-next-qatar

[428] "The Next Qatar?", The Economist, July 27, 2013, retrieved August 5, 2013, http://www.economist.com/news/business/21582272-cost-exploiting-australias-new-found-gas-supplies-soaring-next-qatar

[429] "Energy In Australia 2012", Bureau of Resources and Energy Economics,  February 2012, http://www.bree.gov.au/documents/publications/energy-in-aust/energy-in-australia-2012.pdf

[430] Bruce Leslie, "Solar Thermal Power The Next Resources Boom" Presented to the Australian Institute of Energy, October 25, 2011, available at http://aie.org.au/StaticContent%5CImages%5CBRI111025_Presentation_1.pdf

[431] ReneSola, 156 Series Monocrystalline Solar Module, Irradiation efficiencies vary from 15.9% to 16.2%, available at http://www.renesola.com

[432] Martin, James, "Solar PV Price Index–July 2013", Solar Choice, July 4, 2013

[433] Martin, James, "Solar PV Price Index–July 2013", Solar Choice, July 4, 2013.

[434] "Risk Quantification and Risk Management in Renewable Energy Projects",  AltTran, Arthur D. Little for International Energy Agency (IEA), June 14, 2011

[435] Nussbaum, Alex, "Radioactive Waste Booms With Fracking as New Rules Mulled", Bloomberg, Apr 16, 2014. http://www.bloomberg.com/news/2014-04-15/radioactive-waste-booms-with-oil-as-new-rules-mulled.html

[436] Andresen, Tino, "German Utilities Hammered in Market Favoring Renewables", Bloomberg, August 12, 2013. http://www.bloomberg.com/news/2013-08-11/german-utilities-hammered-in-market-favoring-renewables.html

[437] Wikipedia contributors, "Exemptions for Hydraulic Fracturing Under United States Federal Law", Wikipedia, the Free Encyclopedia, http://en.wikipedia.org/wiki/Exemptions_for_hydraulic_fracturing_under_United_States_federal_law

[438] "Fracking and Water Consumption", SourceWatch Center for Media and Democracy, http://www.sourcewatch.org/index.php/Fracking_and_water_consumption

[439] "Fracking", SourceWatch Center for Media and Democracy, http://www.sourcewatch.org/index.php/Fracking

[440] Schrope, Mark, "Fracking Outpaces Science on Its Impact", Yale Environment 360, http://environment.yale.edu/envy/stories/fracking-outpaces-science-on-its-impact

[441] Kennedy, Will, "Exxon Charged With Illegally Dumping Waste in Pennsylvania", Bloomberg, September 11, 2013, http://www.bloomberg.com/news/2013-09-11/exxon-charged-with-illegally-dumping-waste-water-in-pennsylvania.html

[442] Martin, Allen, "Oil Company Caught Illegally Dumping Fracking Discharge In Central Valley", CBS SF Bay Area, November 26, 2013, http://sanfrancisco.cbslocal.com/2013/11/26/oil-company-caught-illegally-dumping-fracking-discharge-in-central-valley/

[443] "Golden Rules for a Golden Age of Gas", International Energy Agency (IEA), 2012, http://www.worldenergyoutlook.org/media/weowebsite/2012/goldenrules/WEO2012_GoldenRulesReport.pdf

[444] International Energy Agency, "World Energy Outlook 2012 – Executive Summary", available at www.worldenergyoutlook.org

[445] "Golden Rules for a Golden Age of Gas", International Energy Agency (IEA), 2012,

[446] "World Energy Outlook 2012 – Executive Summary", International Energy Agency,

[447] "World Energy Outlook 2012 – Executive Summary", International Energy Agency,

[448] "Honeywell Green Jet Fuel™ Powers First-Ever Transatlantic Biofuel Flight", Honeywell company press release, June 18,2011, http://honeywell.com/News/Pages/Honeywell-Green-Jet-Fuel-Powers-First-Ever-Transatlantic-Biofuel-Flight.aspx

[449] DiMugno, Laura, "Wind Energy Jobs Surface at Presidential Debates", North American Wind Power, Oct 18, 2012, http://www.nawindpower.com/e107_plugins/content/content.php?content.10553

[450] "Renewable and Alternative Fuels", U.S. Energy Information Agency, http://www.eia.gov/renewable/

[451] Miller, Ron, "What the Death of the Sun Will Look Like", April 9, 2013, http://io9.com/what-the-death-of-the-sun-will-look-like-471796727

[452] "Towards a Sustainable Future for All – Directions for the World Bank Group's Energy Sector", World Bank Group, retrieved July 28, 2013, http://www-wds.worldbank.org/external/default/WDSContentServer/WDSP/IB/2013/07/17/000456286_20130717103746/Rendered/PDF/795970SST0SecM00box377380B00PUBLIC0.pdf

[453] Winnie Gerbens-Leenesa,1, Arjen Y. Hoekstraa, and Theo H. van der Meer, "The Water Footprint of Bioenergy," Department of Water Engineering and Management and Laboratory of Thermal Engineering, University of Twente, Enschede, The Netherlands April 2, 2009

[454] Wikipedia contributors, "Olympic Size Swimming Pool", Wikipedia, the Free Encyclopedia, http://en.wikipedia.org/wiki/Olympic-size_swimming_pool, retrieved 28 June 2011

[455] "Cornel Emeritus Professor David Pimentel", Cornell University, http://vivo.cornell.edu/display/individual5774

[456] United States Geological Services USGS, "Estimated Use of Water in the United States in 2005", http://pubs.usgs.gov/circ/1344/pdf/c1344.pdf

[457] "Carbon Disclosure Project Electric Utilities – Building Business Resilience Inevitable Climate Change", IBM, 2009

[458] Fischetti, Mark, "How Much Water Do Nations Consume?", Scientific American, May 21, 2012, http://www.scientificamerican.com/article.cfm?id=graphic-science-how-much-water-nations-consume

[459] Fischetti, Mark, "How Much Water Do Nations Consume?", Scientific American, May 21, 2012

[460] "Fracking and Water Consumption", SourceWatch Center for Media and Democracy, http://www.sourcewatch.org/index.php/Fracking_and_water_consumption

[461] Wikipedia contributors, "Ogallala Aquifer", Wikipedia, the Free Encyclopedia, retrieved 13 Apr 2009, http://en.wikipedia.org/w/index.php?title=Ogallala_Aquifer.

[462] Jane Braxton Little, "The Ogallala Aquifer: Saving a Vital U.S. Water Source", Scientific American, March 30, 2009. Downloaded from http://www.scientificamerican.com/article.cfm?id=the-ogallala-aquifer

[463] Wikipedia contributors, "Ogallala Aquifer." Wikipedia, the Free Encyclopedia,

[464] "Compare Gulfstream G450 – G550 – G650", Aviation News Channel, http://www.decartsnews.com/compare-gulfstream-g450-g550-g650/

[465] Wikipedia contributors, "List of United States Cities by Population", Wikipedia, the Free Encyclopedia, retrieved 24 July 2011, http://en.wikipedia.org/wiki/List_of_United_States_cities_by_population

[466] "Lufthansa 747 operates first transatlantic biofuel flight to US", Aviation Brief, January 16, 2012, http://www.aviationbrief.com/?p=5591

[467] "Technical Characteristics – Boeing 747-400", Boeing company website, http://www.boeing.com/boeing/commercial/747family/pf/pf_400_prod.page?

[468] "Water Consumption", Water for Africa Institute, http://www.water-for-africa.org/en/water-consumption.html

[469] Wikipedia contributors, "Biofuel", Wikipedia, the Free Encyclopedia, http://en.wikipedia.org/wiki/Biofuel

[470] "Jatropha – The Plant", Jatro website, http://www.jatrofuels.com/164-0-Jatropha.html

[471] Bullis, Kevin, "Biofuels Companies Drop Biomass and Turn to Natural Gas", MIT Technology Review, October, 2012, http://www.technologyreview.com/news/506561/biofuels-companies-drop-biomass-and-turn-to-natural-gas/

[472] Bullis, Kevin, "Biofuels Companies Drop Biomass and Turn to Natural Gas", MIT Technology Review, October, 2012

[473] Bullis, Kevin, "BP Plant Cancellation Darkens Cellulosic Ethanol's Future", MIT Technology Review, November 2, 2012, http://www.technologyreview.com/news/506666/bp-plant-cancellation-darkens-cellulosic-ethanols-future/

[474] Downing, Louise, and Gismatullin, Eduard, "Biofuel Investments at Seven-Year Low as BP Blames Cost", Bloomberg, July 8, 2013. http://www.bloomberg.com/news/2013-07-07/biofuel-investments-at-seven-year-low-as-bp-blames-cost.html

[475] Seba, Tony, "Solar Trillions- 7 Market and Investment Opportunities in the Emerging Clean Energy Economy", 2010

[476] Webber, Michael, "Energy versus Water: Solving Both Crises Together", Scientific American, September 1, 2008 http://www.sciam.com/article.cfm?id=the-future-of-fuel

[477] "USDA Offers Loans for Biorefineries", Iowa Energy Center, October 23, 2013. http://www.iowaenergycenter.org/2013/10/usda-offers-loans-for-advanced-biorefineries/

[478] "Combined Heat & Power (CHP) solar generator based on High Concentrated Photovoltaic (HCPV)", Suncore company website, http://www.suncorepv.com/index.php?m=content&c=in dex&a=lists&catid=125

[479] Seba, Tony, "Solar Trillions- 7 Market and Investment Opportunities in the Emerging Clean Energy Economy", 2010

[480] http://www.telegraph.co.uk/news/worldnews/asia/india/10180463/India-sends-its-last-telegram.-Stop.html

[481] Volcovici, Valerie, "World Bank plans to limit financing of coal-fired power plants", Reuters, June 26, 2013, http://www.reuters.com/article/2013/06/27/usa-climate-world-bank-idUSL2N0F300W20130627

[482] "Towards a Sustainable Future for All – Directions for the World Bank Group's Energy Sector", World Bank Group, retrieved July 28, 2013, http://www-wds.worldbank.org/external/default/WDSContentServer/WDSP/IB/2013/07/17/000456286_20130717103746/Rendered/PDF/795970SST0SecM00box377380B00PUBLIC0.pdf

[483] "European Investment Bank to Stop Financing Coal-Fired Plants", The Guardian, July 24, 2103, http://www.guardian.co.uk/environment/2013/jul/24/eu-coal-power-plants-carbon-emissions-climate

[484] Plumer, Brad, "The U.S. will stop financing coal plants abroad. That's a huge shift.", The Washington Post Wonkblog, June 27, 2013, http://www.washingtonpost.com/blogs/wonkblog/wp/2013/06/27/the-u-s-will-stop-subsidizing-coal-plants-overseas-is-the-world-bank-next/

[485] "Coal at Risk as Global Lenders Drop Financing on Climate", Bloomberg New Energy Finance, August 6, 2013, http://about.bnef.com/bnef-news/coal-at-risk-as-global-lenders-drop-financing-on-climate/

[486] Gruver, Mead, "Powder River Basin Coal Lease Auction Receives No Bids For First Time In Wyoming History", the Huffington Post, August 22, 2013, http://www.huffingtonpost.com/2013/08/22/powder-river-basin-coal_n_3794792.html

[487] Fan, Hugh, "A Credit Analysis For Coal Mining Companies", Seeking Alpha, Jun. 19, 2013, http://seekingalpha.com/article/1509622-a-credit-analysis-for-coal-mining-companies

[488] "Walter Energy Inc", Morningstar, http://quotes.morningstar.com/stock/s?t=WLT

[489] Keenan, Mike, and LaCorte, Joseph, "Q3 2013 Review: Coal", Stowe Global, http://stowe.snetglobalindexes.com/pdf/coal-IndexInsights-20131202.pdf

[490] "Facebook: stock quote and summary data", Nasdaq, http://www.nasdaq.com/symbol/fb

[491] "Google: stock quote and summary data", Nasdaq, http://www.nasdaq.com/symbol/goog

[492] Williams-Derry, Clark, "The Hidden Export Bombshell in Cloud Peak's Financials", SightLine Daily, Sept 23, 2013, http://daily.sightline.org/2013/09/23/the-hidden-export-bombshell-in-cloud-peaks-financials/

[493] "How Old are U.S. Power Plants", U.S. Energy Information Administration, http://www.eia.gov/energy_in_brief/article/age_of_elec_gen.cfm

[494] Energy Perspectives, U.S. Energy Information Administration, www.eia.gov/totalenergy/data/annual/EnergyPerspectives.xls

[495] Kubiszewski, Ida, "Powerplant and Industrial Fuel Use Act of 1978, United States", Encyclopedia of Earth, September 3, 2006, http://www.eoearth.org/view/article/155329/

[496] "Energy Policy Act of 1992", U.S. Energy Information Administration, http://www.eia.gov/oil_gas/natural_gas/analysis_publications/ngmajorleg/enrgypolicy.html

[497] Darian Unger and Howard Herzog, "Comparative Study on Energy R&D Performance: Gas Turbine Case Study", Massachusetts Institute of Technology Energy Laboratory, Prepared for Central Research Institute of Electric Power Industry (CRIEPI), Final Report, August 1998

[498] "How Old are U.S. Power Plants?", U.S. Energy Information Administration, http://www.eia.gov/energy_in_brief/article/age_of_elec_gen.cfm

[499] Doughman, Andrew, "NV Energy to decommission coal plants, shift to gas and renewables", Las Vegas Sun, April 2, 2014, http://www.lasvegassun.com/news/2013/apr/02/nv-energy-decommission-coal-plants-shift-gas-and-r/

[500] "NV Energy Acquisition Information", NV Energy company website, https://www.nvenergy.com/company/acquisition/

[501] Davis, Tina, and Goossens, Ehren, "Buffett Utility Buys $2.5 Billion SunPower Solar Projects", Bloomberg, January 2, 2013, http://www.bloomberg.com/news/2013-01-02/buffett-utility-buys-sunpower-projects-for-2-billion.html

[502] McCarthy, James E., "EPA Standards for Greenhouse Gas Emissions from Power Plants: Many Questions, Some Answers", US Congressional Research Services, November 15, 2013, http://www.fas.org/sgp/crs/misc/R43127.pdf

[503] http://pdf.wri.org/global_coal_risk_assessment.pdf

[504] "Energy Subsidy Reform: Lessons and Implications", International Monetary Fund, January 28, 2013

[505] Yang, Ailun, and Cui, Yiyun, "Global Coal Risk Assessment", World Resources Institute, November, 2012, http://about.bnef.com/bnef-news/coal-at-risk-as-global-lenders-drop-financing-on-climate/

[506] "Annual Energy Review", U.S. Energy Information Administration, http://www.eia.gov/totalenergy/data/annual/showtext.cfm?t=ptb0709

[507] "Coal and Jobs in the United States", SourceWatch Center for Media and Democracy, http://www.sourcewatch.org/index.php?title=Coal_and_jobs_in_the_United_States

[508] "Coal and Jobs in the United States", SourceWatch Center for Media and Democracy.

[509] Pyke, Alan, "Coal Workers Lose Pensions As Execs At Bankrupt Company Get Bonuses", ClimateProgress, May 31, 2013,

[510] "Towards a Sustainable Future for All – Directions for the World Bank Group's Energy Sector", World Bank Group,

[511] "Desperate Measures", The Economist, Oct 10, 2013.

[512] Jun, Ma, and Li, Naomi, "Tackling China's water crisis online", China Dialogue, Sept 21, 2006,

[513] Schneider, Keith, "Bohai Sea Pipeline Could Open China's Northern Coal Fields", Circle of Blue, April 5, 2011, http://www.circleofblue.org/waternews/2011/world/desalinating-the-bohai-sea-transcontinental-pipeline-could-open-chinas-northern-coal-fields/

[514] Schneider, Keith, "China's Other Looming Choke Point: Food Production", Circle of Blue, May 26, 2011, http://www.circleofblue.org/waternews/2011/world/chinas-other-looming-choke-point-food-production/

[515] Luo, Tianvi, Otto, Betsy, and Maddocks, Andrew, "Majority of China's Proposed Coal-Fired Power Plants Located in Water-Stressed Regions", World Resources Institute, August 26, 2013, http://www.wri.org/blog/majority-china%E2%80%99s-proposed-coal-fired-power-plants-located-water-stressed-regions

[516] Luo, Tianvi et al, "Water Risks on the Rise for Three Global Energy Production Hot Spots", World Resources Institute, November 7, 2013, http://www.wri.org/blog/water-risks-rise-three-global-energy-production-hot-spots

[517] "South-North Water Transfer Project", International Rivers, http://www.internationalrivers.org/campaigns/south-north-water-transfer-project

[518] Wikipedia contributors, "North Water Transfer Project", Wikipedia, the Free Encyclopedia, http://en.wikipedia.org/wiki/South%E2%80%93North_Water_Transfer_Project

[519] "Bohai Sea Pipeline Could Open China's Northern Coal Fields", Circle of Blue, April 5, 2011, http://www.circleofblue.org/waternews/2011/world/desalinating-the-bohai-sea-transcontinental-pipeline-could-open-chinas-northern-coal-fields/

[520] "A Bulletin of Status Quo of Desertification and Sandification in China", State Forestry Administration, P.R.China, January 2011, http://www.forestry.gov.cn/uploadfile/main/2011-1/file/2011-1-5-59315b03587b4d7793d5d9c3aae7ca86.pdf

[521] King, Ed, "Desertification crisis affecting 168 countries worldwide, study shows", The Guardian, April 17, 2013, http://www.theguardian.com/environment/2013/apr/17/desertification

[522] "Desertification – A Visual Synthesis", United Nations Convention to Combat Desertification, http://www.unccd.int/Lists/SiteDocumentLibrary/Publications/Desertification-EN.pdf

[523] "Rivers are disappearing in China. Building canals is not the solution", The Economist, Oct 10, 2013, http://www.economist.com/news/leaders/21587789-desperate-measures

[524] "Pollution Forces North China to a Standstill", The Wall Street Journal, Oct 23, 2013

[525] Wikipedia contributors, "Particulates", Wikipedia, the Free Encyclopedia, http://en.wikipedia.org/wiki/Particulates

[526] "Particulate Matter Management in the Bay Area", Saffet Tanrikulu, Ph.D., Bay Area Air Quality Management District San Francisco, CA presented at the 2nd Korea-U.S. Symposium on Air Environment Policies Seoul, ROK November 29-30, 2012 http://www.baaqmd.gov/~/media/Files/Planning%20and%20Research/Research%20and%20Modeling/PM%20Mgt%20in%20SFBA.ashx

[527] "Air Pollution Linked to 1.2 Million Premature Deaths in China", New York Times, April 1, 2013, http://www.nytimes.com/2013/04/02/world/asia/air-pollution-linked-to-1-2-million-deaths-in-china.html

[528] Lavelle, Marianne, "Coal Burning Shortens Lives in China", National Geographic, July 8, 2013, http://news.nationalgeographic.com/news/energy/2013/07/130708-coal-burning-shortens-lives-in-china/

[529] Wong, Edward, "Air Pollution Linked to 1.2 Million Premature Deaths in China", New York Times, April 1, 2013, http://www.nytimes.com/2013/04/02/world/asia/air-pollution-linked-to-1-2-million-deaths-in-china.html

[530] Wikipedia contributors, "Coal", Wikipedia, the Free Encyclopedia, http://en.wikipedia.org/wiki/Coal

[531] "Indian Coal", http://www.indianetzone.com/24/indian_coal.htm

[532] Wikipedia contributors, "Coal", Wikipedia, the Free Encyclopedia, http://en.wikipedia.org/wiki/Coal

[533] King, Ed, "Desertification crisis affecting 168 countries worldwide, study shows", The Guardian, April 17, 2013

[534] "Statement of Charles D. Connor, President and Chief Executive Officer, American Lung Association", American Lung Association, May 12, 2010, http://www.lung.org/press-room/press-releases/statement-of-charles-d.html

[535] Wikipedia contributors, "United States Military Casualties of War", Wikipedia, the Free Encyclopedia, http://en.wikipedia.org/wiki/United_States_military_casualties_of_war

[536] "Coal Costs the U.S. $500 Billion / year in Health, Economic, Environmental Impacts pollution", FastCompany, February 11, 2011, , http://www.fastcompany.com/1727949/coal-costs-us-500-billion-annually-health-economic-environmental-impacts

[537] Schwarts, Ariel, "Coal Costs the U.S. $500 Billion / year in Health, Economic, Environmental Impacts pollution", FastCompany, February 11, 2011,

[538] Williams-Derry, Clark, "The Hidden Export Bombshell in Cloud Peak's Financials", SightLine Daily, Sept 23, 2013,  http://daily.sightline.org/2013/09/23/the-hidden-export-bombshell-in-cloud-peaks-financials/

[539] Goossens, Ehren, and Martin, Christopher, "First Solar: May Sell Solar at Less than Coal", Bloomberg, February 1, 2013,  http://www.bloomberg.com/news/2013-02-01/first-solar-may-sell-cheapest-solar-power-less-than-coal.html

[540] Goossens, Ehren, and Martin, Christopher, "First Solar: May Sell Cheapest Solar Power, Less than Coal", Bloomberg News, February 1, 2013, retrieved July 25, 2013, http://www.bloomberg.com/news/2013-02-01/first-solar-may-sell-cheapest-solar-power-less-than-coal.html

[541] "IN THE MATTER OF EL PASO ELECTRIC COMPANY'S APPLICATION FOR APPROVAL OF A LONG TERM PURCHASE POWER AGREEMENT WITH MACHO SPRINGS SOLAR, LLC", Case No. 12-00386-UT, http://164.64.85.108/infodocs/2013/1/PRS20179845DOC.PDF

[542] Note: for a nice Rube Goldberg gif, check out this URL: http://imgur.com/gallery/QCuGNsd

[543] Channell, Jason, Lam, Timothy, and Pourreza, Shahriar, "Shale & Renewables: A Symbiotic Relationship", Citi Research, Sept 12, 2012